THE FUTURE OF THE SPACE PROGRAM
LARGE CORPORATIONS & SOCIETY

Discussions with 22 Science-Fiction Writers
Conducted by **Jeffrey M. Elliot**

GREAT ISSUES

1

OF THE DAY

R. Reginald

the Borgo Press

San Bernardino, California
MCMLXXXI

For Mary Kirkhart—
Who Devours Life with Eager Appetite and Search

Library of Congress Cataloging in Publication Data:

Main entry under title:

The Future of the space program.

(Great issues of the day ; #1)
1. Astronautics—United States. 2. Corporations—Social aspects—United States. I. Elliot, Jeffrey M. II. Title: Large corporations and society. III. Series.
TL789.8.U5F79   303.4'83   80-17954
ISBN 0-89370-140-8 (cloth, $8.95)
ISBN 0-89370-240-4 (paper, $2.95)   OCLC #6581630

Produced, designed, and published by R. Reginald and Mary A. Burgess at The Borgo Press, P.O. Box 2845, San Bernardino, CA 92406, USA. Printed in the United States of America by Victory Press, San Bernardino, CA. Binding by California Zip Bindery, San Bernardino, CA. Cover design by Michael Pastucha.

First Edition————December, 1981

FOREWORD

It has been said that survival in today's world is a race between awareness and catastrophe. Faced with ever-present crises, both at home and abroad, it is vital that we ask ourselves a number of pertinent searching questions: Where have we been? Where are we now? Where are we headed? Where do we wish to be? At its most basic level, this book is about the future—what it can be, what it should be, what it must be.

Clearly, the world cannot stand still, at least for more than a nanosecond. It is always in the process of becoming. If the past has little influence, the future is thereby closer. The touchstone of the future is the freedom it offers—mind-staggering in its variety of choices. In planning for the future, we should not expect to steer a perfect course, but we can and must steer an approximate course. The task of formulating solutions to our problems is not only extremely complex, but increasingly urgent. It involves discarding many of our old concepts and evolving radical new ones. As Rollo May suggests: "The old ideas are dying, while the new ones haven't yet been born."

While attempting to formulate solutions, we must keep in mind that in each of us, and hence in our institutions, there is a continuous struggle between the certitude of the past and the dubiety of the future. The answer is neither to romanticize the past nor look for simplistic solutions in the present. Rather, the future demands that we think in new ways, act in new ways; that we enlarge our perspectives and broaden our visions. This period of crisis, however, has two faces. The ancient Chinese were well aware of this duality, for their character for "crisis" is made up of two symbols, one meaning danger, the other opportunity.

In many respects, we are at the dawn of a new level of human consciousness. Like the early jet pilots and engineers who hovered fearfully on their side of the sound barrier, we hover fearfully on this side of a new consciousness. Like those pilots and engineers, we must with courage and faith break the barrier; for on the other side the dawn of a limitless new consciousness awaits us, while here we can only live in the shadows of yesterday's past. The choice is ours. The time must be now.

This volume, the first of many to come, focuses on the field of science fiction—a literary genre which has long sought to help readers to face the future more confidently—and perhaps intelligently. As science fiction editor Donald A. Wollheim points out: "The troubles that loom ahead may seem difficult to overcome, but science fiction writers have already imagined various ways in which they could be met, by which they could be surmounted, or by means of which we could survive them. Once a thing can be imagined, it can be done—such is the lesson of science fiction." GREAT ISSUES #1 examines two salient topics—the future of the U.S. space program and the role that science fiction writers can and should play in working with America's major corporations to plan for the future. It brings together the collective judgments and experiences of twenty-two leading science fiction authors, each of whom view the future in different, yet similar ways. Rather than draw definitive pictures of the future, they suggest alternative scenarios based on a broad range of possible events. It is critical, as they note, that we view the future with anticipation and understanding, with expanding expectations and aspirations, and with confidence and determination. Perhaps this book, in some small way, will help to set into motion the creative energies and talents which must be marshalled to build that future, whatever it may be.

4

What explains the present lack of citizen interest in and support for the U.S. space program?

MILDRED DOWNEY BROXON

As one who came to science fiction through an early fascination with astronomy, I could think of no good reason to explain the lack of public support for the space program. The fact, nonetheless remains. I asked a few people. Their responses were interesting and often delivered with great fervor. They included:

1. *As entertainment, the space program is boring. Compared with other television offerings, it carries insufficient entertainment value.*

If true, this speaks poorly of the taste of the American public. Do we need bread and circuses? However, NASA could promote its product more aggressively. I understand they have a publicity department, but have never seen any evidence of it.

Science fiction itself may be partly to blame: a mission in which all goes as planned, with no "plot" (i.e., peril and unexpected occurrences), may seem dull by comparison. Personally, I prefer utter predictability to the drama of Apollo 13. Polar explorer Amundsen is quoted as saying that "adventures happen to the incompetent."

2. *Fear of the unknown. "They" are out there and ready to get "us." Most people's sole exposure to space travel is through grade-B monster flicks. In addition, space travel may threaten established beliefs, especially fundamentalist theology.*

5

My first point applies: you can't have it both ways. But must life be merely a choice between boredom and terror? Good, interesting things also happen. Or are we afraid of fracturing the crystal spheres?

3. *It is elitist. There is no way for the average man-on-the-street to participate, nor can he perceive any direct personal benefits, either in improved products or employment for those outside the aerospace industry.*

This, too, seems to indicate a lack of adequate publicity about space-program spinoffs, even if the concept of basic research may be difficult to sell.

4. *It costs too much. Since NASA is associated with the government, NASA scientists are seen as wasteful and overpaid. People have come to distrust anything associated with the government.*

In a letter some months past printed in *Astronomy* magazine, Julie Woodman pointed out that Americans spend one-and-one-half times as much money each year on pizza as on the space program, and three times as much on cosmetics. Detected welfare fraud amounts to twice the size of the NASA yearly budget. Most people are amazed by this and other statistics. Welfare is traditionally pitted against the space program in "either-or" arguments.

After the Apollo 13 near-disaster, my Public Health Nursing instructor delivered the opinion that we shouldn't waste all that money on space; we should spend it on the cities. I'd been up all night fretting and watching the news, I screamed at her for some fifteen minutes without repeating myself. She may not have been convinced, but she became aware that there were other points of view.

Poul and Karen Anderson, in their short story "Murphy's Hall," treat the paternalistic-welfare-state-versus-exploration theme in a way that may well make the reader weep. (This story can be found most readily in Anderson's collection, *Homeward and Beyond.*)

As to mistrust of the government, perhaps private enterprise would do better; not that many people trust large corporations, either. Most early science fiction writers assumed that space exploration—and exploitation—would be privately sponsored.

5. *Unreality of the experience, as shown in the media. We have all seen far too many high-quality simulations and special effects. People do not believe they are witnessing a real event.*

After Apollo 11 many of my acutely psychotic patients were convinced that the moon landing was a fake, a television simulation. A recent film had the same premise, transferred to Mars, which may reveal something about screenwriters.

On the other hand, I've seen amazingly good simulations of, even, one-sixth gravity. Perhaps we could convince filmmakers to start shooting their space sequences on location.

6. *None of us will ever get out there.*

This is a variation of the "elitist" objection, and, alas, valid. One of my most poignant memories is of a panel at a Science Fiction Writers of America banquet day program, in which several astronauts answered questions from the floor. The audience were men and women who had devoted their lives and made their livings writing about space travel. We all knew we would never experience the reality. We were in the wrong field, and we were too old. The astronauts, on the other hand, tended to be rather blase.

Edmond Hamilton explored aspects of this situation in at least two of his stories: "What's It Like Out There?" (which was, incidentally, the title of the aforementioned panel) and "The Pro," about a science fiction writer's envy of his son the astronaut. I highly recommend both. They can most readily be found in the collection, *The Best of Edmond Hamilton*.

My husband suggests a lottery, the prize being a ride on the space shuttle. This would be safe enough for the man-on-the-street. (I have dibs on the story idea!)

The above have been the opinions of friends and acquaintances. They must have spent some time thinking on the matter, since conversation grew lively. Some of them read science fiction, others do not. Their knowledge of my personal bias may have influenced their answers. (I did, some months back, offer to throw one of them into the lake when she said she couldn't care less about the space program. I live on a houseboat, so this was no idle threat.)

I fear that much of the American public has either lost its sense of wonder—tragic, if so—or its sense of wonder may only be sleeping, and can be awakened. The human race needs pioneers, people who are always striving toward another goal, itching to see what lies over the next horizon. On Earth, the next horizon has already been mapped.

So what to do, besides bemoan and deplore? What magic kiss can waken a sense of wonder? People seem to respond better to fictionalized entertainment than to fact; after the first few launches, television did not bother with live coverage, and there was only cursory mention of the fantastic Jupiter fly-by discoveries.

GORDON R. DICKSON

The major factor that explains the present lack of citizen interest in and support for the U.S. space program is, in a word, apathy. The program, which suffered from popular misunderstanding and a consequent curtailment of funds in earlier years, is now in trouble by being taken too much for granted.

The average citizen is still under the comfortable impression that our program is still level with and comparable to—if not having a healthy lead over—space research in other parts of the world. This is not true. Our hardware is aging and becoming old-fashioned. Our workers are old and getting older, and their training is in some cases growing out of date with newer techniques and information. Above all, the gradual strangulation of necessary funds has reached the point at which a number of the major goals have slipped away out of sight.

In short, we are in my opinion already losing the space race. Not only that, but it would take the best-funded of crash programs a number of years merely to catch up the distance that we have fallen behind, let alone regain our earlier lead. What we are going to be faced with is the training and experiencing of a whole new generation of workers, plus the need possibly to have to scrap a great deal of our hardware and start over again from scratch.

Moreover, we're going to have to change our attitude. Ariane, the Commonwealth rocket, in its economy of design, points the direction in which the future of space development is strongly likely to go. We should recognize this fact and take measures accordingly, before we wake up to find that other cultures besides ours have taken the lead in the economic exploitation of space; and that we no longer have the funds to get back into that competition, which will become more and more vital to our economic welfare as the century draws to its close.

RAYMOND Z. GALLUN

The moon-landings demonstrated that an ancient symbolic impossibility could be done. That was the apex of a drama—a long, wondering, hardly believable effort. But, lo—there it was—a fact. Marvelous! In all seriousness, this was perhaps the highest point of our nation. But thrills aren't meant to last without renewal. So pretty soon some aims of a somewhat different sort began to be discussed. "If they can get to the moon, why can't they...?"—insert whatever verb and object. The "they" instead of "we" detaches the demanding wisher from any serious responsibility or effort in the desired action. Tacit thanks—we'll watch. We don't even have really to think. *They'll* produce.

Usually, the various objectives had nothing to do with space programs, except as in the metaphoric phrase, "pie in the sky." So you've got your moon—even a little of Mars and the other planets. Nice. So now extend yourself some—give us Utopia, easy everything, freedom, happiness, dignity enough to spit in everybody else's eye, brotherhood of man, world-wide or at least nation-wide. Trouble is, the equation mixed incompatibles; a relatively simple engineering solution and a matching to-

8

gether of hardware, however huge and marvelous, are hardly in the same class as remaking human nature.

A lot of what was attempted was long overdue, and in part accomplished well before there were moon rockets. But since the moon landings, the why-don't-theys have gotten overblown and largely unworkable—out of step with what people really are. Seldom the angels for innocent idealists, not quite devils either—just lusty and rather selfish much of the time. But still the inertial beat and rhythm of what has been started goes on, trying to wallow its way toward the better world.

So we have a give-me culture. People want the most for the least. Sometimes I suspect that laziness is subsidized. Vast sums are spent on improving education—with little improvement. The list of wastage on other why-don't-theys that work poorly is very long. So we've got inflation. So treasuries say they're going broke; people still are very well fed, but insist they can't make ends met. So there must be raises. The government people say there must be reduced public expenditures. So what takes the cut among the first?—sure, space programs.

Our species hasn't changed much in the last few thousand years; its main focus is on the belly, food, sex, kids, diversions, comforts, some little old thing to strut and crow about—a gold ring for the nose, a nice car or house—whatever. Maybe raising our eyes sometimes toward the sky and the stars—toward space programs—is also in the latter category. The first big show, there, is over, the shouting is finished, the curtain down. And there's a so-what shrug. But I hope the thrill comes back soon. There's too much romance and mystery and room for productive, exciting, and demanding action out there to just let it be, with indifference. It may be that our survival is even out there too, in our dispersal. Whereas, if not enough of us dream and strive again, I have some doubts and wonderings. If we cease to be a viable, constructive, resourceful people, attuned to the realities of ours and future time, if we become oversoft and complaining, somebody with more vigor and realism may take us over and use us; or if we prove even too useless for that, they may just get rid of us.... Or, instead of burning up in the much-feared nuclear holocaust, we may just putrify from our over-pretty and phony why-don't-theys of perfection.

JAMES E. GUNN

The present lack of citizen interest in and support for the U.S. space program seems to me to stem from a couple of decisions made by NASA and other government leaders early in the space program. The first was to assure citizens that spaceflight was a completely normal, safe enterprise; the consequence of this was the public image of a program filled

9

with All-American astronauts and great rooms of equipment and monitors. The very language of the space program was prosaic: ascent stage, command module, extravehicular activity, lunar orbit insertion, portable life-support system . . . NASA opted for the engineering approach, for prose rather than poetry, and the public got bored.

The second decision was to emphasize the immediate goal of landing a man on the moon. It may have been important to mobilize the necessary support to beat the Russians there, but by the third or fourth landing the public was bored. It would have been better to keep constantly before the public that the moon was only the first step into space, that beyond lay Mars, Jupiter, and the outer planets and beyond that, interstellar space—a bit of poetry and a dream.

ISIDORE HAIBLUM

Why *should* citizens support the space program? Citizens, these days, have their own problems. The cities are falling apart, inflation is chewing up people's savings, health costs are going sky high, and unemployment is edging its way toward a national figure of eight percent. The space program made good copy for journalists, and will probably put Tom Wolfe's new book on the best-seller list. But what has it done for you and me lately? Or any of our relatives?

I think it should also be noted that the derring-do characters who fought their way through the pages of our favorite science fiction magazines in the forties and fifties did not turn up in the actual space program. In real life, knowing how to operate a machine proved more important than the ability to make a quotable quip. If Luke Skywalker had had the quality of a real-life astronaut his picture would have been a flop.

With Ted Kennedy throwing his hat in the ring, the Ayatollah hunting down those who "war with God" in Iran, and our television networks grappling for first place in the ratings, how can a mere space program compete?

Now, if any of those space probes had brought back a genuine alien or two, at least we science fiction writers would have probably felt a bit better about it. As things stand, however, the space program looks as dead as a dodo. And just about as interesting.

JAMES P. HOGAN

"I'm having to siphon the lawnmower to get my car to the gas station, my heating bills would buy Fort Knox, the interest on my mortgage just went out of sight, my company is talking about laying people off, the IRS wants the fillings out of my teeth, my fifteen-year-old daughter smokes pot, dates a communist and thinks she's pregnant, and my doctor says that too much health causes cancer. You wanna know why I'm not curious about the infra-red emission from Saturn!''

The response of the average American of today to the question would probably run something like that.

The problem is, of course, that the man on the street sees nothing in the space program or its aims that will influence the issues closer to home, nearer the heart, and demanding on the pocket; therefore he doesn't find time to think about it too much, or any obvious reason for wanting to. On top of this the citizen in today's society of constant change possesses a relatively short attention span of interest that requires the stimulation of "something different" in the daily diet. To satisfy this need he becomes mentally locked in to and dependent on the offerings dispensed by the news media until his thinking is virtually shaped by short-term topics currently featured in newspapers and on television. At the present time the media are not plugging the space program, presumably because it isn't considered newsworthy right now; therefore the average citizen doesn't think about it. Do the media condition public taste or simply react to it? Well . . . that's a separate question.

Another reason is that things like the energy problem, inflation, Middle-East politics, the SALT treaty, and so on all bear directly on the life of the average citizen and his family, and are therefore of concern to him; nothing that happens to a probe several hundred million miles from Earth is likely to have any direct impact. There was a lot of citizen interest in the space program when Skylab was coming down.

Also, regrettably but predictably, I think the country is still suffering from a degree of disillusionment after the magnificent achievement of the Apollo program. Apollo, of course, was very newsworthy; the media plugged it extensively, and everybody thought about it and sustained interest in it. If science could lay down a detailed schedule for a task as awesome as getting to the moon and then proceed to carry it through calmly, confidently, in detail and ahead of what seemed an impossible deadline, then surely science could do anything. All of the problems and hardships that had plagued mankind since history began were about to be swept away forever. The masses had at last found their panacea . . . or so for a while they thought. In fact, if any power has the potential to wipe away humanity's scourges, it is indeed that of science and reason. But scientific discovery is a slow, steady process that requires perseverance, patience, and tenacity; people who didn't understand science wanted all the results *now*. But years after man landed on the moon people were still starving, nations were still fighting, and Utopia hadn't happened. The children wanted to see their new toy working and grew impatient waiting for daddy to put it together; some went back outside to play while others sulked and decided they didn't want it anyway. But once it's working okay they'll be happy for a while . . . until it becomes just something else to be taken for granted and they get bored. Then they'll be ready for something new.

Finally, the authorities responsible for "selling" and publicizing the space program tended to adopt a somewhat naive approach in the form of stressing the short-term, trivial spinoff benefits such as non-stick frying pans and pocket calculators instead of educating public awareness of the longer-term and deeper implications of the whole thing. In other words, they were pandering to the whims and fancies of childhood instead of recognizing and respecting the potential to grow up. The true significance of a fully fledged space exploration program lies not so much in the spectacle of extraterrestrial activities *per se*, which is exciting enough to be sure, but in the setting up of the industrial base needed to make such a program possible, along with the high technology, advanced research, skilled labor-force, and educational system that go with it. These are the foundations needed upon which to build a society capable of breaking through the problems confronting the human race at its current point in history. The problems are essentially to do with a minority of "haves," too many "have-nots," and "finite" resources.

Resources are finite only within the framework of a given system of economics and technology. In past eras of history the work that could be done by a man's muscles, the number of slaves he could own, the number of horses he could buy, and so on all represented finite resources which limited the amount of wealth (expressed ultimately in terms of provision for food, shelter, clothing, and attainable living standards) that he could create in a year. In the advanced sector of today's world such factors are irrelevant and whether or not they are finite is immaterial; we worry instead about oil, gas, raw materials, deforrestation, and so on. These things are finite within our current economic and technological framework.

Periodically, however, the human race makes quantum leaps into totally new realms of possibilities and achievements that are qualitatively distinct from anything that went before, rather like the phase changes that govern the transitions of a substance between solid, liquid, and gaseous states, where completely new laws of behavior take over and the old laws cease to apply. An example was the discovery and harnessing of electricity, which opened up whole new realms of phenomena that were not simply extrapolations of the technologies of earlier centuries. When such quantum leaps in technology occur, they invariably open up new dimensions of individual opportunity, life style and ultimately freedom, and are usually followed by periods of political and social upheaval while society adapts to and absorbs the new values created. It's interesting to note that the major problems confronting

12

the government of sixteenth-century England were over-population (relative to what the economy of the time could comfortably support), inflation, unemployment, and fuel shortages . . . with a population that was tiny compared to the incomparably better-off one of today.

The infinite extent of space stretching away on every side of us surely symbolizes that there are no limits to the growth of our civilization either in terms of what can be achieved or of how far it can spread. The powers of human intellect and inventiveness have repeatedly broken down barriers to progress previously thought to be insurmountable, and shown themselves capable of creating new sources of wealth, energy, and productivity which have carried us through progressively higher levels of control over our environment.

An energetically supported, comprehensive space program would signify most importantly a commitment to the attitude of mind that must be adopted if the human race is to break out of the straitjacketing acceptance of the need for austerity, scarcity, conservation, and zero-growth. These are imagined limits consequential of our current economic system and the conventional technologies upon which that system is based: they are not absolute. The sciences and technologies required to achieve a realistic program of space exploration and construction—including for example a fully nuclear-based economy worldwide which would lead eventually to controllable fusion plasmas—would afford not merely abundant and cheap power, but also revolutionary methods of primary metal extraction (applicable to lunar rocks), cheap raw materials of all kinds, economic large-scale desalination of sea-water (and hence abundant food through irrigation programs), and inexpensive "cracking" of seawater to yield hydrogen as a basis for the manufacture of cheap synthetic fuels. Given all that plus space travel, the "limits" to growth all crumble away. Just as the pressures that built up in Europe four hundred years ago precipitated the wave of migration and expansion that founded the New World, so those same pressures repeating on a global scale today will trigger the move up off the planet, out of the solar system, and who-knows-where after that.

The average Neanderthal hunter was probably too busy to pay much attention to whomever invented the wheel; the medieval peasant probably didn't care about what the first ocean-going sailing ships would find over the horizon; the midwest farmer at the turn of the century probably didn't get excited about the Wright brothers' experiments at Kitty Hawk. Perhaps if the average citizen of today were allowed to share more of the vision of the tomorrow he is helping to build for his children, their children, and the generations that will follow them instead of being told about frying pans and computerized tic-tac-toe, he'd feel differently about the space program and where it's leading.

13

ROBERT A. W. "DOC" LOWNDES

You don't have to have studied history (though it helps); you only need to have observed people to realize that most of them do not take a long view of anything unless (a) they have been trained to do so or (b) they see a tangible, reasonably certain benefit from doing so or a clear danger in not doing so.

Consider cancer research. That arouses public interest and support because so many people have had close experiences with cancer—a friend or relative had died of it—and they are motivated by a combination of fear and hope. Perhaps next year, if not this year, a cure will be found, and it's understandable that the research needs public support. That is very concrete.

But the U. S. space program, for all its spectacular aspects (highly entertaining), is still relatively abstract. So we landed a man on the moon; so we're getting fantastic photographs of interplanetary space and finding still more moons of Saturn—what has that got to do with the price of potatoes? There was a time when that question was a gag—but not now. The public knows that the price of everything is rising constantly; that is very concrete.

If there are any tangible benefits that space research and the U. S. program has brought to the public—anything that can be seen and experienced in day-to-day living—the word hasn't trickled down. What the public is aware of is that it's a very expensive, government-backed program—and government programs are widely known for wastefulness, corruption, and political ploys. No one runs for office promising great benefits to the public from the U. S. space program, if necessary funds and support are obtained. Who would believe the candidate who made such promises anyway? Whether candidates are charlatans or sincere, they have to run on the price of potatoes, etc. Publicity about the space program may distract the newspaper-reader's or television-viewer's mind from such concrete matters momentarily; but not long enough to win solid support for the program.

As to the dangers to our national security in *not* keeping up with the Soviet Union, etc., on a space program, the public has been trained to believe that warnings about any real danger are made by "wolf-" criers, and put them in the same category as warnings that the Soviet Union really does intend to bury us as soon as it becomes convenient. The American public is free *not* to support the space program (whereas the Soviet public has no such freedom); it can only be *persuaded* to do so, without the type of coercion that you find in other countries.

So far, the propaganda for the necessity of supporting the space program has been less than universally convincing; only a relative few pay any attention. Whether (and when, if) that situation will change remains moot.

14

RICHARD A. LUPOFF

To anyone who grew up loving science fiction, it's appalling—and baffling—to contemplate the lack of interest in and support for the so-called "space program." Why, to those of us who were practically weaned on the old images of the swamps of Venus, the sands of Mars, and the rings of Saturn and the great red spot of Jupiter, it's the most natural and obvious thing imaginable, to pursue that great outward urge.

First, explore.

Then, colonize.

If we meet anybody else out there, it will be the most glorious and most exciting event in history (and perhaps the most dangerous, too, but life *is* risk).

Well, what the astronomers have taught us is that there *are* no swamps on Venus. In fact, the place is about as close an approximation of the classical Hell as one could ask for—even though it was John W. Campbell who told us years ago that the moon was Hell. Nup. Venus.

Mars is a little closer to our classical conception, but it's so damned cold and there's so very little oxygen that it's really unliveable except in a totally artificial environment (e.g., a sort of diving-bell-in-reverse, something like a LEM. Don't ask me what a LEM is or I'll burst into tears).

And so on.

But . . . .

But, we science fiction folk say, so what? The challenge is all the greater. Let's go there anyway, do it moon-landing fashion. Good grief, doesn't anything survive of the spirit that drove Tenzing Norgay and Edmund Hillary up Mt. Everest? Why climb the highest mountain in the world? Because it's there!

Why travel to the planets?

Because we're a bunch of little crawling maggots bound here to the surface of a tiny mudball endlessly circling a minor star out in a boondocky neighborhood of this galaxy—unless we get up off this tiny mudball and go out to meet the universe.

And the planets are only the second step. (The moon was the first.) We might just encounter some wonderful surprises on the moons of Jupiter and Saturn. But what if we don't? They're still stepping stones, and we'd better start moving, and go on, and on, and on forever (or as good as forever, from the temporal perspective of a puny human).

Who ever heard of a caterpillar who preferred creeping on his belly all his life, to soaring as a butterfly? That's what we're doing, it seems to me.

Why still: Why?

Well, I'm afraid that the average human being is vastly more concerned

15

with the immediate and everyday realities of mundane life, than he or she is with such seemingly remote and abstract notions as sending a spaceship to land on Neptune. That's understandable. We all have our worries and concerns, bills to pay, worrisome symptoms to take to the doctor, careers to pursue, kids to put through school, and a thousand and one more matters. Who cares about another planet? Who cares about another star?

It's the President and the Congress who decide how the Federal budget is going to be spent. And no smaller entity than the Federal government—not any single state, not any corporation or foundation or university or voluntary association—can support the effort that it will take to explore the solar system.

And the President and the Congress set their priorities and allocate the resources under their control, so as to gain and hold the favor of their supporters and constituents: corporations, lobbies, institutions, voting groups as perceived along many different bloc criteria (i.e., age groups, economic groups, ethnic groups, geographical groups, etc.), and ultimately "the people" *en masse* and as individuals.

Well, parents care about their kids' education, old people care about retirement benefits, sick or injured people and potentially sick or injured people (which is to say, everybody) care about medical research and treatment. People who drive care about streets and highways and people who fly care about aviation safety and convenience.

And so on, and so on, and so on.

And of course we're all scared that the Russians are going to invade us. Also the Chinese. And the Cubans. And the Iranians. And the Cambodians (if any). Whoever.

And they're all afraid of being invaded by us and/or one another.

Which leads to the hackneyed but nonetheless relevant point that the world armaments race gobbles up hundreds of billions of dollars worth of resources every year. If and when we can stop that madness, there will be a "peace dividend" not only for the U. S. but for every nation in the world.

What will become of those resources? Not that any such happy problem is likely to occur, but just in case . . . what will become of those resources? Can we get the President and the Congress—not to concern ourselves at the moment with other nations—to devote those resources to no less a goal than the exploration of the universe?

Think about that. Roll that phrase around on your brain, and on your tongue. Don't be afraid of it. Don't be embarrassed: "the exploration of the universe."

But the President and the Congress will be under many other pressures, very *legitimate* pressures, as for how to devote the resources. Medical research and treatment facilities, urban renewal, job training and development, educational advancement at every level from pre-school to post-doctoral, replacement of ruined housing and upgrading of salvageable housing, feeding the world's hungry mouths, and so on *ad infinitum*. How can a remote and abstract concept like "the exploration of the universe" compare with something as real and immediate as filling empty bellies?

Well, maybe we can fill the empty bellies *and* explore the universe. And maybe the two things can interweave, and in the process of travelling to the planets, and eventually to the stars, we can learn things and find things that *will* feed bellies, cure cancers, build houses, and so on and on.

In the process of developing the Apollo capabilities that brought us the moon, a great many useful and practical inventions and discoveries were made. Or so I've been told. NASA and the other interested parties certainly did a lousy job of publicizing those achievements. Maybe a good public relations campaign is what we basically need.

Or maybe we could just let the three television networks bid for exclusive rights to footage of the mission to Jupiter. That might take care of paying the bills right there.

LARRY NIVEN

Do I really have to tell this audience that most of humanity is mundane?

My family were all highly intelligent people. A couple of my uncles read *Analog*, but most never read science fiction. *One* of my high school teachers did, the rest didn't. My physics teacher at Cal Tech hated science fiction (and still does; we met again recently). I didn't discover the Los Angeles Science Fantasy Society until I was twenty-five. Until then I simply accepted the fact: most people don't care whether someone lands on the moon or not. Most people don't try to predict or shape the future except in the most personal and local fashion.

And I didn't *know* any idiots. These were the bright ones.

So now we've explored the moon. It cost us a tiny percentage of the cost of the welfare budget or the Vietnam war. I never heard anyone complaining about the taxes for that project . . . except for quotes in the newspapers, from congressmen. Maybe the percentage of mundanes in Congress is denser than that of the general population.

We've explored the moon, and there are still mundanes. Why is everybody surprised? I never was.

My collaborator Jerry Pournelle has harsh words for the NASA publicity department, whose peculiar skills allowed them to write of the most exciting event in human history, and make it dull. Reverse Midas Touch, he calls it. Me, I get irritated when I remember how many science fiction

17

writers attended the launchings at Canaveral, by scrounging press passes. What should have happened is this: the newsmen who attended should have been forced to prove that they had published at least one science fiction story within the previous five years.

But these are symptoms of a basic fact: most of the world doesn't care. They put their effort into shaping their own futures, assuming that what is true now will go on forever. The percentage of *us* is much higher than it used to be—thanks to the lunar landings and their live television coverage, thanks to those mind-blasting pictures from Jupiter and Saturn, thanks to *Star Trek* and *Star Wars* too—but it's still small.

What to do about it?

1) Don't panic. *The mundanes don't stop us.* They don't care either way. The most they'll stir themselves to is blocking tax dollars from funding interesting research. So—

2) Stop buying Wisconsin cheese until Wisconsin fires Senator Proxmire. Find other targets among those who block government funding for research in space, for orbitting power plants, for the shuttle and for alternatives to the shuttle.

3) Be careful who you vote for. (Vice-President Mondale made his position clear in 1976. He thought the shuttle was the most important decision then to be made, and he was against it. The shuttle budget was chopped immediately after the election.)

4) If anyone asks, you're in favor of the neutron bomb. (If any of us survived that war, we'd still have a civilization to return to. It's not certain that civilization could be rebuilt from scratch; the easy resources are already gone.)

5) Find out who your congressman is. Write him expressing your views. (People have been telling me that forever; I'm sick of hearing it, but it's still true.)

6) Push where you think something will give. Tell heart patients where their monitoring equipment came from. (It's *all* from the space program, from instruments designed to tell ground-based doctors what was happening to apes and men in orbit. My Dad got an extra twelve years after his first heart attack, because of the space program.) Pass the best (and most nearly set in the present) of your pro-space novels on to your mundane friends and relatives.

7) Don't give up. Knowledge doesn't get lost. Today it spreads at lightspeed. Every defeat is only a postponement; every victory is permanent.

Thirty years from now, the mundanes will think that the value of the orbitting solar power plants was obvious to everyone. And they'll be moaning about the tax dollars going into a permanent space station outside Jupiter's bow shock wave.

CHARLES SHEFFIELD

Is there really a lack of interest in the space program? I think there is not. I think the average citizen finds the *results* of the space program fascinating. My evidence? First, the number of articles that are written on such things as the Voyager fly-by. I do not think that you would see such intensive coverage if people were not interested—commercial interests would prevent it, since the media have not usually been motivated largely by altruistic and uplifting feelings. Second, look at the Air and Space Museum in Washington, D.C. It outdraws all other museums there by a factor of at least two, and it is the most-visited museum in the whole world. People are fascinated by the results of our air and space exploration. Third, talk to the astronauts who have given round-the-country lectures. They will tell you that the first few questions they get in an evening may be on the costs and the benefits of space exploration. But after that first few minutes—almost as though it were needed to satisfy people's consciences—the question of cost disappears, and people begin to ask "What's it like out there? What would it be like for *me* if I were able to go out there?" That's what turns people on. I think there are millions of people who groan when the latest TV program about the 1980 election comes on, but who will sit and listen and watch programs on space and its mysteries.

But what about the question of *support* for space activities? Now we are into a complex area. The question of space program support is usually asked not as a stand-alone thought, but more as a question of *alternatives*. If you go to someone and say, "Are you willing to give up your TV football coverage (brought to them, as often as not, over communications satellites) in order to fund the Galileo mission?," they'll answer—and who can blame them?—"Hell, no. Anyway, what's the Galileo mission?" If you try and take money away for a purpose that people don't understand, it is not surprising if they don't go along with it.

Why don't they know what the NASA missions are? That's a sad situation. The public relations that NASA conducts about its own programs is unbelievably bad.

The result of all this was stated long ago by Robert Heinlein: "NASA has taken the most exciting development in the history of the human race, and made it *dull*." I like to think of it as the Reverse Philosopher's Stone—they turned gold into lead.

So, *interest* in space, yes, it's there in abundance. *Support* for space expenses? Not the way it is being sold now—I wouldn't buy it myself, if that's all I knew.

But suppose that you broaden the question and go out and ask people, "What does the U. S. have that is its most treasured possession?" A

19

few may talk about the Constitution, or democracy, but a surprising number will tell you that it's *American know-how* that separates us from most of the rest of the world. This country has had a long history of being first in science and technology. Look at the number of Nobel Prizes that we gather each year in the sciences—it's more than any other nation on earth, comfortably so, far more than the Russians are awarded.

So if you were to say to people, "Are you willing to spend your tax dollars in developing American science and technology?," you'll get a lot of support—even from the same people who are scared of, for instance, nuclear energy development. There is a basic understanding that we depend in our lives more and more on an effective technology, even though there may be resentment of that fact.

Now, how does science and technology relate to the space program? Over the past twenty years, it has been the tip of the pyramid, the lead edge—pioneering in medical work, in micro-processing, in materials, in remote measuring, in monitoring of processes—you name it, and during the sixties and seventies, if it involved research work, the space program has made a direct (usually) or an indirect (more rarely) contribution.

Why is there this apparent contradiction, that people sense the need for us to spend our money more and more in science and technology, to put ourselves in the forefront of development there, but at the same time they do not think we can afford to spend their dollars on space? Obviously, they do not see the connection between the two. And obviously, someone is failing to draw the line of connection clearly enough.

Who do I blame? Basically, I blame the scientists themselves, for failing in two different ways. First, there is a degree of compartmentalization now in science, one of narrow disciplines and limited scope. This is caused partly by the fact that we need to specialize more to understand any field—*all* fields have been growing at a fierce rate—and partly by the fact that the word "scientist" does not mean what it meant a hundred or even fifty years ago. Now, perhaps eighty percent of practicing scientists are "career" scientists, who expect a good living to come from their efforts, and who realize that the luxuries will come only if they are not merely successful, but *famous* (i.e., *provenly* successful). With such attitudes, scientists regard their territory as jealously as any starling, and encourage the tight compartments that true science should deplore. These same compartments will now compete for funds—not agreeing that a dollar spent on the possible discovery of new particles may offer the final tool for cancer cures eighty years from now.

If I am right, modern scientists have become their own worst enemy, by failing to recognize and emphasize the inter-connected nature of all science.

I said the scientists had failed in two ways. The other failing arises from specialization also. I prefer to call it bad public relations, but it is just as validly a failure to realize that science cannot prosper without a set of common objectives. Instead of the message being "Solar power is better than nuclear power," or "Cancer work is more important than solid state physics," or "Put money into laser research and not into artificial intelligence," the united statement put forth by all scientists, regardless of their discipline, ought to be, "Investment in science and technology is the most important investment that the U.S. can make."

Think of almost any group in this country (e.g., the military, the medical profession, the mine workers, and so on) and you will find that when it comes to issues in their common interests they will speak with one voice. Visit Washington, any day of the week, and you will find on Capitol Hill a large number of lobbyists for the oil companies, for the insurance companies, for the banking interests. But you will look for a long time before you find the "science lobby." It is considered as something that is *beneath* scientists—and perhaps it should be. On the other hand, unless the equivalent of a science lobby is set up, I do not think that we will ever see Congress moving forward to give the country the unified science program that most people admit we will need in the next twenty-five years.

Is anything being done about that? Not much, but let me mention something that in a small way shows what I believe will be the pattern for the eighties. In the past few months, the major professional societies interested in space have begun a series of discussions on cooperative efforts. Recently, on behalf of the American Astronautical Society, I met with representatives from the American Institute of Aeronautics and Astronautics, the L-5 Society, and the Aerospace Industries Association. The object: a loose federation of these and other societies, which will identify common goals, and initiate programs to move toward them with the combined resources of all the member societies. This is a long way from the idea of a single science lobby, but I think it may be the first hint of something.

It's interesting to wonder why this should begin to happen now, at the very end of the seventies. It ought to have happened during the *early* seventies, as soon as it became clear that the space program for this country lacked definition and purpose. Ever since 1969, there has been no really new goal for space.

I think that the seventies were a period when we did not quite *realize* that we had reached the end of one piece of the road. It takes a time for people to admit that the dream has gone sour—especially when we had so much in this country to keep us occupied. In the past two years, we have begun to wake up again from the long sleep, and wonder where the eighties will take us. The simple answer is, if we don't know where we *want* to go, chances are we'll finish up someplace else. As our inability to control our use of oil and other resources becomes more painfully obvious, the small fraction of the country who do any long-range thinking are worrying about our options. Those options are hard to evaluate—but it certainly looks as though the next ten years will prove again that we are as strong and safe a country as our science and technology permits us to be. We have lost some ground to the Japanese and the Russians in the past seven or eight years, but in most areas of lead-edge technology we are still the front-runners. We have to organize to reap the benefits from that front position, and to make sure that we stay there.

I don't mean we will be interested in using our science in an oppressive way. I mean that our balance of payments, and our whole success in the international sphere, will be dictated by our possession of tools and techniques for science that are still not controlled in the rest of the world.

The space program is vitally important to America. It is one of the key pacing items for science and technology development, for the production of the advanced methods that decide the world position of this country (and the internal standard of living) for the next ten to twenty years. That's how the space program has to be presented to the public—not as an isolated series of space spectaculars, not even as the bread-and-butter efforts of the weather satellites and the communications and earth resources satellites. It will be the single most important item in deciding the position of this country on the international scene. The spin-offs will come in both the civilian and the military areas, and the need for more advanced technology is there for both. (It would be nice if we could drop the military efforts—but at the moment we know for a certainty that the Soviet expenditures in that area keep growing and growing.)

Let me put all this together. What explains the lack of public interest and support for space? Simply this, that the public does not know the relation of space to all science and technology. Why doesn't the public know it? Because we have not made the point clearly enough, loudly enough, and unitedly enough.

I have hopes that this will change in the near future, and that we will create a "space lobby." It will not be funded out of profits, like the milk, oil, banking, and wheat lobbies. It will instead be funded at a meager level, and will obtain its effectiveness because of a shared belief of a very large number of people.

JACK VANCE

As I view it, the question is best answered by five basic points:

1) Public taste has been satiated and corrupted by *Star Treks* and horror movies; the space program by contrast comes up drab, listless, and lacking in human relevance. No monsters, no girls in gossamer night-

gowns, no evil dictators.

2) Space-program personnel in general and the astronauts in particular are insipid to the point of anonymity: over-trained, over-disciplined stencils.

3) The public quite reasonably is bored by the sterility of our near neighbors: Mercury, a cinder; Venus, the same; the moon, an airless desert; Mars, nearly so; and Jupiter, incomprehensible. No interesting landscapes or jungles, not to mention intelligence. Photographs from space, such as those of Jupiter and its satellites, indeed are magnificent, but unreal.

4) The time scope of star-travel and even planetary exploration is discouraging. When and if hyper-speed space-travel becomes real, and star-travel can be accomplished in weeks and months, rather than aeons (hopefully without Einsteinian distortions); when an adventurer can take off in his private star-boat and personally participate in star explorations, then surely there will be no dearth of public interest. Absolutely and certainly to the contrary.

5) Certain elements of society complain of "money thrown into space." Nothing, of course, could be further from the truth. The value of the spaceship is not intrinsic; it derives from human work, for which people are paid. Space money is spent on Earth, funding a new industry which eventually will offer careers to millions of persons now "disadvantaged."

A. E. VAN VOGT

I have my own theory about how the space program came about in the first place. It was not John F. Kennedy who was behind it. You may recall that John was an avid reader of "James Bond"; that was his escape literature, so to speak. Robert—it turns out—read science fiction. I believe the older Kennedy sponsored the space program for his beloved younger brother.

All this came upon an American public that had no thought about such matters, except—and here is a basic characteristic of Americans—they were beginning to react to Sputnik's soaring ride into solar space.

The mere idea that Russia had got there before the United States evoked a peculiar disturbance. It was an anxiety response. The people of this country have grown up with an in-bred feeling that we are the technological giant of this planet, superior to all those lesser types out there. It was this automatic, continuing ego trip that received a shock from Sputnik's achievement.

So no one objected to the necessary multi-millions being spent. Even after John F. suffered his sad personal disaster, the project went on almost automatically. There was no resistance from Congress, and shrewd President Johnson built all the expensive buildings and installed all the machinery in the South, with his own state, Texas, getting the prize control building. People seemed to recognize that that was politics, southern style, and no one was concerned.

And so when the time came there we all sat glued to our television sets, watching the moon landing with an interest that varied for the individual; but essentially the average person's curiosity was quickly satisfied. An extreme example of the general reaction was that of the father of a fellow writer—who shall be nameless. The father took one long look at the desolate moon landscape, and said in total disgust, "For God's sake, it's nothing but a desert!" Whereupon he switched over to his favorite television program. And that was his absolute final interest in the space program.

What all this tells us is that Americans are a "now" people. During the oil crisis there were the long lines of cars, the owners of which were out to get more than their share. Meaning, they were making doubly sure that they wouldn't go short. Since the shortage lasted longer than the first moon landing, the reaction spilled over into two separate areas, both of which profoundly affected the economy. One: there was a great rush to buy small cars for the good mileage they gave; this is still going on, as is evidenced by the fact that the price of the imported Honda has a fifteen hundred dollar tag on it over and above the original retail price. Honda is an example of many similar price increases on small cars.

Response number two: instead of selling their old machines and buying new ones, millions of Americans decided to make do; and so they are now overwhelming service and repair shops, paying high prices for the poor workmanship being delivered by the hastily-trained mechanics needed to take care of the rush.

One result: the beautiful big models gathered dust on ten thousand lots; and the ruination of all the Detroit manufacturers was under way—beginning with Chrysler.

What this tells us about the future of the space program goes something like this: it will proceed quietly at its present low level, insofar as American financial support is concerned. (I should take a moment here to acknowledge an unknown factor: other countries, such as West Germany, have quietly financed special space experiments in which their scientists were interested. Such outside financing, if it grew, could play a role of importance.)

Congress is actually spending more than the American people want them to, but since it's done quietly, no one objects. It would probably be ideal if the money needed could be included in the military budget. If that could happen, the project could go forward with only an occasional announcement from Cape Kennedy, such as: "Yesterday, the Air Force launched three Saturns carrying thirty-seven astronauts on a new moon project." And that would be the only news item. No further information, except that a year later there would be pictures in *Scientific American.*

That possibility has one strike against it. The military budget is already under attack; and top generals would resist money being diverted away from the routine colossal things that they feel they need just to maintain basic security.

So the actual way that our space program may get into gear again will probably be old style. Suddenly, we will learn that the Russians have established a permanent base on the moon, called "Brezchnevgrad." The shock will spread more slowly this time, but it will build up. And, presently, we shall have a moon base also. Once it's there it will receive the necessary financing to maintain it. And the subsequent discovery of gold diamond mines will feed back the reassuring thought that the thing will now pay for itself. The lulled feeling will continue indefinitely; and the subject of how much it all costs will never again be of interest to anyone other than a few kooks. The expense of the space program could become one cent added onto the price of gasoline. Sold by the liter, of course.

You may say, why can't we assume that Americans are capable of planning for a future space program? After all, they buy personal insurance for their own future. My answer: no, they don't buy it; they are *sold* insurance, and they contribute to social security without choice.

We need to remember that our democracy is a great experiment in dealing with human nature as it was observed to be. In its fashion, democracy will presently take us to the stars. But that fashion, when subsequently examined, will look strange, indeed. However, once it happens, no one will be able to examine it properly. And so this that you read here is the only objective examination you may ever see.

One more thought: it should be noted that democracy has never postulated a change in basic human nature. Marxist theory does; at some remote time all communists will be perfect beings. Meanwhile, the place is run like a prison; and things get done if somebody decides to do them. Our hope is that somebody will.

JOAN D. VINGE

I can think of two possible causes for a falling-off of citizen interest in the space program. One is inherent in the program itself, unfortunately— it has been in existence for long enough now that it has become "old hat" to much of the public. The days of a competition with the Russians for dominance in space have passed; John F. Kennedy's goal of putting a man on the moon by 1970 was achieved a decade ago. The projects that NASA is concentrating on now are potentially much more useful—projects like the spaceprobe exploration of the solar system and the development of the space shuttle—but their visual and media attractiveness is generally much lower than that of the early space flights. This is primarily because the time span required to show results from projects such as the Jupiter fly-by is measured in years—long years during which nothing is heard about its progress. People simply forget about it in the meantime. And too, there are no glamorous personalities associated with an unmanned probe, or with the shuttle program. When there were only seven astronauts, we all knew their names. But now there are hundreds of astronauts and other scientists waiting to ride the shuttles—all of them virtually anonymous.

However, I think an even greater reason for the erosion of public interest lies in the current state of society itself. The past decade has been one which has seen the boom economy of the sixties shrivel, while inflation, unemployment, and the energy crisis have focused peoples' attention strongly on their own comfort and survival. The long range vision required for an enthusiastic support of the space program has shriveled up along with the Affluent Society; government spending on projects not seen as directly affecting "down to Earth" necessities of life are regarded as inflationary and extravagant. Coupled with this narrower forcus is a basic disillusionment growing out of the social unrest of the late sixties. The mistrust of technology that was a part of the sixties' feeling, combined with a seventies' sense of futility about our ability to change or even control our environment, has probably caused many people to feel the old belief that "technology will cure all ills" is no longer true.

Whatever the reason, it is true that the magic has gone out of the space program for many Americans. I think this is a very unfortunate situation, because the space program aside from giving us amazing glimpses of what lies beyond our sky, has also given us the beginnings of a number of important industries—from transistors and microprocessors to solar energy. The potential for future industrialization in space is high, once we pass the threshold of the space shuttle program. The U. S. space program is by no means a cure for all our ills, but it can offer us ever increasing benefits if we continue to give it the opportunity.

CHELSEA QUINN YARBRO

It seems to me that there are several factors contributing to the apathetic, not to say hostile, attitude of the general public to and for the space program. First, and perhaps the most difficult factor, is the increasing distrust of government, combined with an almost universal dismay at the increase in government spending. In the wake of Watergate and other such scandals, the average voter is unwilling to trust those in power. While this skepticism is, on the whole, a healthy

26

sign, it means that long range projects, which the space program most certainly is, will be regarded with a great deal of suspicion and lack of enthusiasm.

Second is the aforementioned time element. Most people in this country are taught from childhood that there must be quick visible results to projects and activities or the work is not worth doing. The space program does not give instant, easily visible results, and therefore there are a great many people who are simply not willing to wait the requisite number of years to see *if* it was all worthwhile. Many of the benefits that NASA might demonstrate are somewhat ephemeral to the average citizen, which contributes to the problem.

Third, the pervasive climate of anti-intellectualism in the country makes it difficult for any project to gain general support if its most obvious proponents are members of the intellectual community. To make this worse, most of the visible members of NASA, other than the astronauts themselves, are both military personnel and engineers. Between the fear of scientists, who brought us the H-bomb, and engineers working with the military for what are feared to be nefarious ends, most citizens fear the implications of the space program, real or imagined.

NASA itself has contributed to this hostility. For one thing, in their eagerness to make the space program seem less dangerous than it is, they also robbed it of most of its excitement. A great many people regard going to the moon as an expensive sort of milk run. It is difficult to maintain support for a program that is costly and dull.

Last, and in some ways, most important, is the conviction that most citizens have that they have not been told everything, that the space program is a front for something more unattractive, or is being used for purposes other than the stated ones. I must admit that to some degree I share this opinion. The cult of secrecy that has surrounded much of the space program contributes to this feeling. If NASA is indeed doing more than what it appears they are, it would be wise to be a bit more open with the public. Rather than point to teflon pans and medical equipment, it might be a good idea if NASA could say that they are curious to discover if we have any company in this part of the galaxy, or mention the various mysterious phenomena that they can investigate only from space.

One of the things that many highly-placed men of the space program seem to have forgotten is that most people are curious. Deny them the pleasures of curiosity with high-handed treatment or the patronizing you-would-not-understand attitude, and those having to support the project will come to resent the whole space program. A greater degree of candor combined with a realistic presentation of time-tables and goals might go a long way to restore interest, enthusiasm, and confidence in the space program and its various projects.

There are two factors, as I see it, the first one being the current state of the economy, causing the administration and Congress to be more cost-conscious. This has made it more difficult for NASA, with a shrinking budget, to set up long-term goals and, hence, to make predictions of achievements in space. A realizable space time-table would help in focusing public attention on the space program. We simply do not have one which arouses any great interest. For example, no date has yet been projected for a permanent manned space station.

The second thing seems to be that we do not look upon the space program as representing national strength in the same way that the Soviets do. They have sent twenty-six cosmonauts into space in the past four years, to break all of our Skylab records and to log half again as many man-hours as we have overall. And they talk about it with obvious pride, as a demonstration of their technical achievements, as a matter of international prestige.

If you want to know what I think is going to happen, I believe that the Soviets are currently working on a space shuttle of their own and will have a permanent manned station in space within the next three years—crude, perhaps, compared to the best that we could manage with full funding—but something capable of being expanded, something which could eventually serve as the basis for assembling in orbit a manned interplanetary vehicle. I feel that it will be the military possibilities of such a station, however, that will act as a spur to our own space program, much as the first Sputnik did in 1957, causing us to step up our own efforts—probably with a crash program—to set up a permanent manned space station of our own. At such a time, there will probably be more public interest and support, as it comes to be looked upon both as a matter of defense and prestige.

What role can and should science fiction writers play in working with America's major corporations in planning for society's future?

POUL ANDERSON

Offhand, I doubt that there *can* be any such role. While individual executives may occasionally be interested in science fiction, the big corporations are run by bureaucracies with little or no more collective imagination (or guts) than the government. Companies do nowadays consult "respectable" futurological outfits such as the Rand Corporation or the Hudson Institute, and these may in turn solicit the opinions of a few writers on the q.t. But isn't that about the maximum extent of it?

I'm not even sure that anything further would be desirable. Science fiction writers have no more of a pipeline to the future than does anyone else, and are apt to have less exact information than specialists do. We brag a lot about having used such themes as atomic theory and spaceflight from the start, but I don't know of a single story which came anywhere near matching the course of real events in those areas. And did any writer foresee, say, the transistor, with all its implications? On the contrary, our characters were still using slide rules, hundreds of years from now!

Does science fiction have any social value, then, beyond being entertainment and perhaps, in a small way, helping keep a degree of literacy alive in this country? Well, yes, I would say it does. Among other things, by its own multifariousness it keeps showing that the future is *not* fore-

29

seeable, is certain to be full of surprises. As Ben Bova has remarked, it can play with wild cards—for example, contact with an extraterrestrial civilization—that are simply not allowed to the professional futurologists. Thus, in however shadowy a fashion, it does sketch out alternative scenarios, and keep its readers aware that predictions are seldom fulfilled.

Of course, this does not mean that careful, professional extrapolations should not be made and taken into account by those responsible for the planning of important undertakings. In fact, one of the few encouraging things I see these days is that the directors of public and private organizations are doing so, rather than drifting along quite blindly.

In a small and usually indirect way, science fiction can help keep them alert to the fact that surprises are inevitable, and thus perhaps give them a little extra flexibility in their planning.

I think, though, that by far its most significant service to society is among the young, recruiting the scientists and technologists we so badly need, combatting complacency and, at the same time, opening eyes to the fact that there are far more hopeful prospects for mankind than the Barry Commoners and Jane Fondas of this world let on.

Maybe some of those young readers will get into positions of responsibility before it is too late. Maybe.

OCTAVIA E. BUTLER

I'm not sure most science fiction writers are taken seriously enough to play any but an indirect role in influencing corporations in particular or society in general. At any rate, the indirect role is the one I will comment on.

The work of science fiction writers is writing. This does not mean their influence is limited to readers or, in the case of film, to viewers, but it does mean they must influence readers and viewers first, and influence them strongly if they are to reach anyone else. So, then, what role should science fiction writers play in helping their readers and viewers (young people in particular, whether they eventually become corporate executives, blue collar workers, scientists, whatever) cope with societal change and plan for the future?

Science fiction writers should provide a kind of Madison Avenue for unfamiliar or "unacceptable" ideas. That is, ideas that need to be considered, gotten used to, perhaps adopted, but at least judged with as little as possible of fear, prejudice, ignorance, or that natural human conservatism which causes people to suspect or reject the unfamiliar automatically.

Science fiction writers can "advertise" unfamiliar ideas by including them within good entertaining stories. For instance, most science fiction

30

writers use the metric system in their stories of the future. I doubt that this has put every science fiction reader at ease with metrics, but I haven't heard the hostility toward a metric change-over from science fiction readers that I've heard from non-science fiction people ("unnecessary," "un-American," "communistic").

Familiarizing people with unfamiliar ideas eliminates at least some of the apparent alienness of those ideas. And familiarizing people with unfamiliar people eliminates at least some of the apparent alienness of those people. Science fiction has long done this with people who might or might not exist—extraterrestrials. They've done it so well that some nonfiction writers have had an easy time convincing large numbers of people that the extraterrestrials are already among us everywhere—and everywhen. Unfortunately, many of the same science fiction writers who started us thinking about the possibility of extraterrestrial life did nothing to make us think about here-at-home human variation—women, blacks, Indians, Asians, Hispanics, etc. In science fiction of not many years ago, such people did not exist, existed only occasionally as oddities, or existed as stereotypes.

Happily, this situation has improved, especially for women, but science fiction is still not doing nearly the job it could do, and do naturally without preaching, without proselytizing. There are still people in the genre who think nothing should change. For instance, at a recent convention a writer, a professional, explained that he had intended to use a black character in a story of his, but decided not to do it because he felt that a black would change the focus of his story, make it a story of race relations. His thinking was too narrow for him to visualize a black in any other context. Moments later he found a way to eliminate blacks from even race-relations stories. He suggested blacks be represented by extra-terrestrials.

On the opposite side, though, the side of the kind of advertising I'm advocating, are writers like Suzy McKee Charnas (*Motherlines*) and John Varley ("The Persistence of Vision") who portray blacks and members of other groups as exactly what they are—members of the societies in which they live. People. Neither writer focused on race relations as a theme, and neither seemed in any danger of being sidetracked by their black characters. They just told their stories—Varley well enough to win a Nebula. Good writing. Good advertising. It can be done, should be done in the hope of making some small impact on the retrogressive, fragmented present as well as on a future that could be better.

C. J. CHERRYH

Corporations don't consult with writers about anything. If the world is

31

lucky, corporation policy-makers *read*, and in that sense, you could call it consulting. If the corporation policy-makers are imaginative, they read both history and science fiction. The present society does too much of its thinking and planning in ten-year gulps—as if a decade could tell us all we need to know about the world. Bunk. Civilization is 10,000 plus or minus years old on this planet. Our sun is a mote of a star spinning along like a speck of dust in a limb of our galaxy, and we're a mote beside that mote of a star. *That's* reality.

Readers of science fiction are people with the Long View. The energy crisis? We predicted it long ago. The solutions? We've written about those too. We don't make up the laws of nature—we just make up the stories that use them . . . and machinery on the drawing board today; and we wonder, good, creative wonder. We ask common sense questions and moral questions and technical questions. We what-if anything and everything that's proposed or possible, and show how it might affect living things. We're the bridge between the sciences and the humanities, the machine and the human question. We ask why, why not, and what next; and when the future looks threatening we find ways that can take us through it in good order. The energy crisis—we've solved it. We know the answer. We're moving on. By the time the world gets there, we'll be further still.

Science fiction is for people who can see things as they *really* are . . . .
And with luck—the planners of our future *read*.

In some cases they haven't—but not so strange to say—nature has a way of getting one's attention in the long run. And we look back at those living the present which was the future when we first saw it, and say . . . umn—hmn . . . told you so.

Two propositions.

First proposition: our academic institutions, particularly in humanities, have fostered a degree of myopia in the young . . . and by two deadly processes.

One, the reaction of the humanities to the machine age. When James Watt's steam engine proved practical, when the Industrial Revolution started . . . literature branched in several directions, one branch tending in the way of defeatism and the studied opinion that the machine would be the death of mankind; another branch tending toward romanticism of the machine—Verne, Wells, etc.; yet another tending toward romanticism in the past or adventure avoiding moral issues or statements of attitude altogether, which might be subdivided into various other categories. Generally, the academicians fell into the camp of the anti-romantics and praised the so-named "vision" of the purveyors of catastrophe in the human soul; they ignored the historical and the adventure fiction as trivial, concocting terms such as "popular" fiction and other such labels to set it in its place. But those romantics who could see hope for

32

mankind's future, who could believe in romanticism in a world which contained machines . . . why, said the academicians, they must be depraved and worse.

Science fiction from its origins as a separate art form met with a chilly reception from academe—because science fiction and the anti-romantics are natural enemies, and the anti-romantics were well entrenched in the literary establishment and the philosophy departments (the members of which, after all, were not engineers). So the Approved Reading List and the prize winners were generally those so-named visionaries who feared the machine, whose vision of reality was limited to the decade, and whose heroes (a misnomer) were as small-sighted.

So academe worked diligently to amputate vision, and hope, and to convince the young, who are born to hope, that it was all out of fashion. Never mind that the machine had meanwhile failed to destroy us, that the overall lot of mankind on Earth was improved, that we could provide luxuries for the poorest which ancient *kings* would have envied, that the world drew into a village in which our potential for understanding matched our potential for destruction, in which advances in transportation made possible the alleviation of human suffering by shifting surplus to areas of need, in which we have made our first tentative reachings into the wonders of the universe . . . Narrow your vision, my son; romanticism is out of fashion.

Two, the rejection of the classics. After all, now that we have cut off the future, how can the past remain relevant? Why subject students to *The Three Musketeers* or the *Iliad* or to *Robin Hood* or *Hamlet*? Romantics all. Down with 'em. We give them the books of the decade. We make everything Relevant . . . to the decade. The past doesn't matter, my son; we've eliminated it; after all, they didn't have the opinions we all hold Now.

So academe reared us to fear the future, to cling trembling to this narrow little strip of time we call reality, to distrust vision of past or future.

Second proposition: in an age in which science and the humanities sometimes look to have diverged one from the other, this is not the case with science fiction—which from its beginnings has been posed between the two camps.

We suddenly find ourselves short of energy and (surprise!) there is a roar of anguish from the Common Man. We need our machines for the standard of living we've learned to want. And while we have the occasional myopic back-to-natureite who wants to send us all back to the woods again (can you imagine what would happen to the forest with all the former city-dwellers hacking it up for firewood?) . . . mostly we have a gut-deep conviction on the part of the man on the street that We'll Find a Way. Romanticism rears its head in the most unlikely of places. In fact, we *will* make it, barring someone's unthinkable stupidity, barring myopia on the part of our leaders . . . and perhaps in spite of it all. The machine turns out to be important to the man who used to sneer at Buck Rogers, who sits with a pacemaker in his chest, watching satellite weather on television in the twilight of his life . . . what next, for his son and daughter?

The humanities and sciences have to be wedded together again. That's what science fiction does. That's why in one part I rejoice to see schools waking up and putting science fiction into their curriculum; and in the other I shudder . . . to have science fiction crammed into a mold meant to teach Literature, analyzed for form and structure and all of that . . . and robbed of its essence, which is what-if and why-not, and if-this-goes-on. It is a philosophy as much as a literature, an attitude and a way of looking at the whole universe. In the medieval university there was a department of study which was supposed to weld the humanities and the sciences together: Philosophy, called the queen of sciences, helping a student to unify what he had learned in a way which our compartmentalized education does not do . . . we fear to teach philosophy in our schools, because we have no commonly agreed ground for it, no common experience, and because it gets into controversy . . . .

But we have the medium.

That's where science fiction belongs in the curriculum, in a halfway point between science and the literature course, a forum for discussion, for argument, for free debate and for linking past to future . . . the meeting-ground of the history department, the science department, and the literature department.

It has to do with widening vision, with seeing relationships between knowledge and ethics, between hypothesis and belief, between the seen and the equally valid unseen worlds.

That's where we have hope of catching the attention of the corporation. Ah, we have infiltrated: many a young person in the computer sciences, in the technical end of things . . . is a science fiction reader—and ideas filter up, in the best organizations, as well as down. The saddest thing is that the person who goes into business via business education so often walks a corridor apart from the sciences and the humanities, and never manages to find the time in the requirements of his chosen major to take many courses in either area. He may read, may choose these things on his own; may instead content himself with the ten-year segment of so-named reality palmed off upon him by his casual contact with the humanities—may even believe in it.

The capacity to dream. Regarding corporations, and the state of affairs, and science fiction, and whether we science fiction writers are ever consulted . . . the capacity to dream is the real question. Some give it up, mistake that for maturity, and their own decade for reality; and some keep it, and ideas are born of it.

34

Forward-looking corporations are aware how rapidly apparently un-related fields of endeavor may be knit together by some technical ad-vance. In an age when we have trained a plethora of tunnel-vision spec-ialists, there is also room for the generalist, and the inquisitive mind. Many a corporation has discovered that there may be some bright and innovative ideas in the head of the young computer specialist who reads science fiction; and even on the part of the high school scientist who came out of left field with an inspiration he or she conceived by asking what-if.

We have our ways.

Now what might be done . . .

First, I'm not holding my breath until General Motors calls me up and *asks* me my opinion on the future of transportation. I have one. Many of us do. But somehow the direct lines are not there. It probably would not occur to a large corporation to consult with a fiction-writing generalist— for legal reasons regarding the ownership of patents, I should imagine; and for organization reasons regarding in-house personnel who are supposed to be doing the advising; and because someone would have to propose it and that's risk. But we're generalists; we cross the lines of disciplines, and innovation is our profession. We do think the unthink-able, but we don't live divorced from current events. Get a group of us together and you have academically qualified experts in linguistics, chemistry, sociology, history, fine arts, astronomy, engineering . . . but who speak a common language and who understand each other perfect-ly well; get a panel of us together and pose us a question and ideas flow apace. That's one thing that goes on at science fiction conventions, that keeps an audience sitting there firing questions at the panel long after the hotel air conditioning has collapsed from the strain and programs are fanning furiously . . . everyone could leave. They don't. Ideas are hap-pening. Things are being born. An oceanographer in the audience is on his feet making an observation to the linguist and the engineer who are writers, and people around them are understanding. The chemist and the sociologist who are writers reach for the microphone, having gotten an idea at the same moment and take turns expressing their inspiration. This is the kind of thing that goes on at the convention the company technician may have spent his weekend attending . . . on his own time.

I wonder if some corporations might not benefit by calling in such a group, loosening the ties and the preconceptions, listening a time to the writers talk to each other and then throw it open to the audience, no seniority, no structure, no labelled experts . . . no in-field jargon. Just good minds enjoying themselves at full stretch, in an atmosphere in which there's neither status nor right-wrong nor schedule. Frightening, per-haps. Threatening to order, perhaps. As long as a corporation needs assembly-line thinking, each in his isolated cubicle, well, that works for some. And there's a time for that . . . when you have the idea, and it's just a matter of working at it. But there's a time for creativity. And out in that convention audience at the hotel, in the unbuttoned collar and the casual slacks, the fiftyish gentleman listening to the oceanographer and the sociologist . . . he's a corporation president; and the fellow by him with the patched jeans and the tee shirt . . . a practicing attorney; the young woman with the slogan tee shirt is a university instructor; the older one at her right the editor of a national magazine; the fellow next over is a clerk in a store; and his wife, mother of one; and the next young man works for the government. It's that kind of mixing that helps provide the chemistry . . . where you never tend to ask what the man you're talking to *does* or *is*; you judge him by the quality of what he's saying and the sense he makes. By common consensus, *nothing* is counted but a man's presence, nothing is despised but a closed mind, and anyone is free to speak to anybody at all.

Not likely really that a corporation could recreate that.

But those kinds of meetings are going on, in nearly every weekend in the year, somewhere in the United States of America and outside it.

And the doors are open.

RAYMOND Z. GALLUN

Science fiction scribes and Big Corporations? Put them side by side—in imagination—and we have an odd wedding. But an interesting one. Cooperation or contention? And how much of either?

Right away we run into two stereotypes: the Dreamer who can do *anything* in his head—though, often enough, he is apt to be all thumbs. And the Practical Man who knows his nuts and bolts, and has to sweat, struggle, figure the problems out, dirty his hands, and somehow *make the Thing real.*

Between these two legends, there is ample room for an exchange of sneers: that arrogant, ivory-tower idiot! That dumb, primitive grease-monkey—no imagination!

Put the science fiction writers in the first category, and the Big Corporations in the second.

Stereotypes have the flaw of being generalizations—both in, and out of, contact with Reality to some extent. But they *do* have their uses as rough tools for comparison purposes.

One thing that especially delights me is reading a yarn by a science fiction scribe who obviously knows his shop-language, and the difficulties that can come up in a factory. He can dream, but he has been where-it's-at, too, intimately. He smells of how-it-is.

Just as much, it delights me to notice a science fiction magazine on an engineer's desk, or on the bench of an assembly technician. I've enjoyed bits of both these discoveries. So there's a hint of harmonious blend. A mutual crossing-over. An understanding and cooperation. Engineers— even many businessmen—have imaginations, too.

That science fiction writer is interested enough—though often only from economic pressures?—to find out about the other side, do the other fella's job a while, and learn something he perhaps had little idea of. My impression is that he ought to be a better writer from the experience.

The corporation engineer who reads science fiction—and it seems a long-standing fact that they exist!—proves that he finds sufficient interest in its ideas, at least for diversion. He may chuckle in derogatory amusement, still, perhaps subliminally, some fragment of what he has read may stick as an idea-germ. And this may be more especially true if the writer knows, as well as he can, what he is talking about.

So much for useful, progressive harmony. Not that I would ever urge total harmony! That would be unreal and quite unhuman, and perhaps ultimately not very constructive—since contention, at the debate-level anyway, should be part of the means of sifting and checking factors of any proposed action toward viable decisions. Besides, arguing is fun—I can think of several science fiction scribes who seem to make a major sport of it!

However, at some point, we have to expand our view to include the Big Society, in which we all have to live, and somehow—by whatever compromise—must find some place for our much smaller selves to exist, without too much buffetting and being knocked about. Lots of people will surely say that it's a crazy mess. Politicians contending for votes, that they may have some command—likely for some kooky but appealing scheme for betterment which can make things worse. Banners of one kind or another raised by loudly approving groups—only to be shouted down by others.

So here we all are—the Body of that Big Society—surely various and contentious in our individual selves, but in the total mass—what? A huge, amoebic lump, slow moving and primitive, not altogether either bright or stupid, but with a vast inertia, that if started in one course, tends to keep on, on a path that may not only have become obsolescent but downright backward. Though, after a while, it can be prodded around to flow in some other direction, which can prove just as muddled.

Barring the introduction of some mind-science—the overall effects of which I would feel dubious about myself—this great *powerful*, slowly pliable bulk of people is not likely to change its nature very much, very soon—except, hopefully, to be less scared. Folks have always wanted first, much the same simple things that other animals need: food, home, love, sex, family, security, not too much work; enough excitement to keep interested. The rest is individual—or minorly collective—window-dressing, with some good points to it, too.

37

So where do the dreamer—sometimes a science fiction scribe—and the doer—sometimes the corporation-man—fit in all this? What should their roles be? Some science fiction writers might want to take on the cloaks of messiahs—follow me—I have found the Word of Reason—or of some God?—and will show you the way to Better Things! . . . Or a corporation man might do the same. With both, it is a monotonous arrogance, which has fallen on its nose often enough in the past. Yet, as in the past, it remains part of the human game. I don't doubt very much that there will always be these assured persons—no doubt now and then accomplishing something worthwhile. Likely, followers are as eternal.

But both dreamer and doer remain part of the human mass, the overall average opinion of which remains the controlling factor, entirely valid or not. Which means that the individual, though he may try to influence, has to compromise with Public and Family, and try to give them what they seem to want.

Let me wander a bit here, to bring up points of background, as they seem to me: the Big Corporations have become a more or less constant target for blame by politicos. I am sure that it is a partly valid accusation; nothing composed of a large collection of various persons could ever be guiltless. But more, the Big Corporations are convenient bad-guy dummies to pummel—they're huge, and the little guy out there feels small and put-down and powerless, by comparison; he resents them; to a large extent, he is easy to persuade that they are evil. But he isn't told— not very loudly, anyway—that the Big Corporations aren't necessarily monster, blood-sucking entities, but are composed of thousands of people—employees with fairly good jobs and quite similar to himself, or that the financial aspect consists very largely of likewise small folks who have invested their savings, and are drawing fair dividends. So, to a degree, the innocent, angry little critic has been hoodwinked. Indeed, the Big Corporations are neither angelic nor models of total efficiency, but they do successfully manage to fill a massive amount of human needs . . . Some people may shout for Government Take-over—Government perhaps meaning in their minds, some white-knight image, as virtuous, fierce, and just as idealized as a portrait of Uncle Sam. In truth, Government is just another bunch of bureaucrats—folks like others, as straight or crooked—but generally less experienced than corporation executives in their particular fields, and perhaps as selfishly oriented toward an agreeable salary, which they expect to get, regardless of profits, or lack of them. Myself, I think I much prefer the corporation man, since his selfishness is more likely to be harnessed in a constructive direction; if his company and its stockholders don't make money, the word will get around, and he will be out of a job.

Of course, a certain amount of spitball firing at the Big Corporations is inevitable and desirable; careless and destructive practices should be noticed, pointed out, and corrected. But, if avoidable, it seems to me the criticism shouldn't be as vicious and obstructive as it sometimes has been—conjunctive with the emotional snap-judgments of various groups that get stirred up. I refer to things like the Three Mile Island mishap, in which—no doubt partly by luck—nobody even got provocably hurt. The Big Corporations may be justly blamed for hurried and slipshod design and construction procedures. On the other hand, I suspect that the footdragging indolence and inattention of government-bureaucrat inspectors may be at least just as guilty; if they had been really doing their jobs with efficient promptness and care, the incident might not have happened. And in further defense of the Big Corporations, let me add that, according to some reports, the delay in granting construction and operation permits has been rather disgraceful, tying up equipment and construction crews and expanding costs while they dawdle. No wonder the Big Corporations hurry at the end, and get careless. As for the inspectors' and politicians' motives, there might be a bit of the good old human urge to show personal power. The rest of it probably goes back to worry and caution about what ourselves—our huge, sluggish mass of populace—may think and get scared over. Of course fission energy is dangerous—as any high concentration of energy is! But if we say we need the power to run our gadgetry, the answer should be to institute practices of care—as has been done before—since fission energy seems the only effective source of additional power immediately at hand.

Somewhere, here, we come to the question of how the populace is moved, and by whom. Of course one can cite the Madison Avenue Method, and the vociferations of pressure groups, not in the least forgetting corporation lobbies. But at a certain point, it may be that the whole thing becomes more unpredictable than weather.

No doubt science fiction writers are participants—quite possibly I'm one, myself, right now—for good or bad, though, by my best judgment, with good intentions! We may try to be minor guiding lights—to the extent of registering some opinion, or favoring certain actions we think should happen—in the speculative stories we write. Or, similarly, dramatizing what we think are the defects of some prospective idea. All this, of course, is as it should be.

However, we are not free. Like the corporations and the politicians, we are often bent by what the public mass out there seems to want. In the thirties, forties and fifties, it seems we were a lot freer, first with F. Orlin Tremaine's "Thought-Variant" stories in *Astounding*, and subsequently in John W. Campbell's quest for original ideas. Fortunately, then, we were on the right course to attract a large number of forward-looking readers. Much of this seems to me to have changed.

Lip-service is still given to originality. But try to practice it, as was once possible and welcomed, and some editor will get nervous. He is looking at his book-sales sheets—as no doubt he must—since profits are the survival-blood of a publisher. He has to try to guess, what types of stories—on the basis of past evidence—will sell in large quantities. There he comes into contact with the tastes and choices of what seems now the larger bulk of readers, today. So he concludes—perhaps with regret and some trepidation—that the Big Public wants More-of-the-Same. The result is a large bundle of cliche plots and yarns—localed way out there some place among the stars—pseudo-medieval setting—magicians, dragons, etc.

I have nothing against this genre—fantasy, really, loosely bound, too, under the heading of science fiction. And I certainly have nothing against J. R. R. Tolkien, whose yarns I have much enjoyed. Nor do I object to escape-literature, as long as one foot is kept squarely on the ground. What bothers me is the apparent emphasis on escape, and in a repetitious manner. So the reader wants some light reading, which doesn't require much thought or study? Maybe that's it? . . . But isn't this clear evidence of a Future Shocked society, not entirely willing, even, to face the present, and taking refuge in mysticism?

I recently dropped in on X-Con in Milwaukee. There, I am sure nobody told me—I am still not convinced that I heard it right!—though the guy was certainly knowledgeable!—that, according to some reliable poll, only four percent of science fiction readers favor actual human extension into space! To me, this seems incredible, and goes cross-grain to what I've always thought was the right direction. And I believed that science fiction was a forward-looking medium. What has happened? What do these guys want—retreat into the flea-bitten Middle Ages? Or don't they know it was plague-and-misery ridden? What are they trying to do—shelter other worlds forever from the philistine boots of modern man? We may not be perfect, but we do have our virtues. We are at least *Real*—as our dreams can be real, and even, perchance, rather magnificent.

The above is, of course, just my opinion. Whether the portion of the public that we can influence a little, can or should be turned in the direction I favor, is another matter.

I suspect that some of the Big Corporations might—at least for profit-motives—be on the side of my view. Likewise, surely some of the science fiction scribes.

I suppose it is conceivable that the Big Corporations might set up some arrangement—if we are talking about cooperation between them and the writers to make the public more future-oriented—whereby they would finance the publication of literature aimed at brightening the aspect of that objective.

But the idea of another propaganda-program of this sort—when there have been so many others—sticks a little crosswise in the craw! More arm-twisting by Special Interests?

I suppose what will—and perhaps should—happen, is that those scribes who are interested in this kind of future-orientation, will go on trying to put it over as best they can. Their function is not to command or to coerce, but to suggest. Hopefully—little by little—from a combination of causes and nudges, the people will get used to Now and What Might Be—adopt a less shook up and scared view, and move toward the Future with some optimism and enjoyment. And the Big Corporations with them.

JAMES P. HOGAN

What role can science fiction writers play? That of taking a long-term view of the future of humanity and defining the vital role that corporations must play in helping to shape that future. In doing so, writers can contribute to promoting a sense of purpose and motivation that go beyond the necessary but shorter-term considerations of profitability and corporate survival.

All too frequently the problems confronting the world today are seen merely as distorted, parochial issues that have to do with things like heating water for the bathtub or keeping the living room warm in winter. Such issues are trivial and inconsequential. The underlying problem is of a far vaster, global scale and revolves essentially around the question: are resources finite or not, and who gets what share? Two-thirds of the human race does not generate sufficient income per capita to enjoy reasonable standards of nutrition, health, education, and so on. The advanced-sector populations are able to live the way they do because the labor that earns their wealth is performed by machines (i.e., their industrial productivity is multiplied thousands of times over). A high level of productivity can be supported only by ample supplies of energy, and when we get away from the bathtub-thermostat way of thinking and express the problem in terms of the demands per capita of a whole planet advanced to today's Western standards, that means *lots* of energy. Energy on such a scale can be provided only by harnessing to the utmost extent possible the creativity, inventiveness, and intellectual potential unique to the human species; in other words, within a framework of positive economic growth, heavy capital investment, high technology, and a massive program of technology transfer to the Third World. The solution does not lie in conservation, plateauing out into stagnation, or going back to spinning wheels and windmills.

The notion that energy is finite and is going to run out one day is absurd, but the mentality that allows such convictions to take root is positively dangerous. If the human race allows itself to believe that resources are limited, then ultimately people will squabble over them and then fight over them—all the way down to the last barrel of oil. Eventually, when the shortages become acute enough, very probably such a war would escalate and become nuclear. So when you think it through, the only final alternative to nuclear reactors is nuclear bombs; which would the environmentalists want?

I worked for a large computer corporation that began twenty years ago with three men and a capital loan of eighty thousand dollars; today the company employs around fifty thousand people worldwide and has an annual sales revenue of well over one billion dollars. If, on the day the company was founded, somebody had accurately forecast the demands for expenditure that would arise over the next twenty years and asked the president how he was going to meet that out of eighty thousand dollars, the answer would have been, of course, that the suggestion was ridiculous; the company would invest its starting capital and create the income needed to pay its way.

In the same way, we can regard tiny pockets of fossil fuels and easily accessible metal ores that happen to be trapped around the Earth's surface as our starting capital to float the "business." To attempt to stretch them out forever would be ridiculous. We use those as the investment to build up the technological and industrial base necessary to develop the higher forms of energy-yielding methods such as nuclear fission and fusion, and to get off the planet and eventually out of the solar system to realms of virtually inexhaustible resources of every description, thus opening up limitless possibilities. If that move is not made when the time is ripe, and a zero-growth, conservation policy is allowed to take hold, then eventually that starting capital will be used up and civilization will collapse all the way back to the Stone Age. After that it could never again rise since the fuel and other resources needed to make the first step would be gone. Maybe across the galaxy many emergent civilizations have reached that point and gone into decline because they ran out of courage and vision at this critical juncture. Einstein showed that the physical universe must be dynamic and either expanding or contracting; it can't exist in a static condition. It's the same with society in the long term.

A more way-out thought follows from the prevalent belief among scientists today that there are almost certainly many more worlds inhabited by intelligent species, and that sooner or later we will collide with some of them. The history of evolution and the vastness of the universe teaches us in unmistakable terms that nature does not have any "favored children"; with successive advances in scientific knowledge, first the Earth, then the sun, and now even the whole galaxy have been deposed from their privileged positions as centers of the universe and rendered totally insignificant. The lesson has been spelled out time and time again that a species that allows itself to drift into stagnation is ruthlessly extinguished when it encounters a more advanced kind. Which would we prefer? To gain a niche in the survival stakes, you have to get better all the time; those that are overtaken are destined for oblivion. The human race must gain its niche by its own efforts and its own abilities; there are no benign "big brothers" in the sky watching over us to lend a friendly hand or magical, mystical agencies operating to solve our problems for us.

The future of humanity therefore hinges on its confidence in science, technology, and national thought as the only solution to its problems, and in establishing this confidence the corporations can play a leading role. This is especially true in these days of disenchantment with political leadership. Political leaders are too sensitive to the whims of electors, and electoral whims have shown themselves to be the most fanatical and least informed elements in society. The corporations can provide a desperately needed sense of purpose and direction by standing for and identifying with our longer-term destiny. Be profitable and survive, of course, because that's what the interests of both shareholders and employees depend on, but let that survival be seen as a means of contributing to a greater end that has meaning rather than an end in itself.

Why? What's in it for the corporations? Traditionally, the American has been motivated by the feeling that the job he does, his life, his work, and his existence had a purpose beyond a weekly paycheck to pay the rent; he was part of an outfit that was going somewhere that meant something. That was what transformed this country in a couple of centuries. It seemed to come to an end with Apollo. Since then people have come to feel a sense of purposelessness and frustration. Many corporations today are experiencing problems with motivation, finding that the traditional carrots and sticks don't work as they once did; more and more, status symbols and prestige attract mediocrity and leave the talented indifferent or cynical. But the feeling of contributing to something that *matters* is still one of the strongest motivators known, and genuine enthusiasm can only result in better salaries, higher profits, and improved satisfaction all-around. Thus, serving the long-term, more idealistic need would have a direct and beneficial effect on more immediate goals too, which has to be the basis of a good deal.

RICHARD A. LUPOFF

Having spent a couple of years in a rather civilian-dominated army headquarters job, and then having spent twelve years in a business environment where "cutting edge" technology was an everyday concern (i.e., computer development), I'm surprised and pleased that this question has to do with influencing major corporations rather than government. From all that I've seen, government is not only monstrously inefficient and slow-moving, but hopelessly *short-sighted*, and what we need is *long-sighted* leadership.

Not that I want to paint all of business in sparkling colors, either. There are more examples of heartlessness, greed, inhumanity, and both narrowness and shortness of vision in the business world than you can shake a printout at. Just a couple are the use of environment-damaging and cumulative insecticides, and the stubborn clinging to environmentally harmful and non-renewable fuels when clean and self-renewing power sources are readily available and requiring only investment and development to become commercially practical.

In 1933, Frank K. Kelly had a story in *Amazing* ("Into the Meteorite Orbit") predicting that the African nations would become a dominant bloc by covering vast regions with solar collectors and offering the power to an energy-hungry world when fossil reserves ran low!

People are always repeating that science fiction predicted this or that technological development, from the submarine and the airplane to nuclear power, television, computers, and organ transplantation. What is less often noted is the fact that science fiction has predicted every major world problem of our century: overpopulation, pollution, energy-shortage, nuclear accidents, declining literacy, deteriorating quality of life, and so on.

The problem is, nobody would listen!

A secondary problem is that, even if people in positions of influence *wanted* to listen, they would hear not a clear voice but an incoherent babble predicting everything from utopia to disaster, laying out uncountable problems and innumerable—even conflicting—solutions to those problems.

What I'm saying is, even if the president of General Motors, no less the President of the United States, said to the science fiction community, "All right, lay out for me the biggest problem that you see for the next twenty-five years, and the best and most practical solution to it,"—there would be fifty or five hundred answers! How would that president know which one or ones were of value?

There is the "new science of futurism" that we hear of nowadays. Perhaps I'm being unfair, but I must say that what I've seen of the "futurists," especially of their publications, has *not* been very impressive. More drab academic publications are not what we need.

It thus looks like a hopeless situation—but strangely enough, this problem which we seem unable to solve (giving the chief executive officer of each of the *Fortune 500* corporations a free subscription to *Galaxy* won't accomplish much, believe me!) may not need us to solve it. The problem may be solving itself.

Science fiction has made major inroads on college campuses in recent years. Where "that nutty kid who reads Buck Rogers stuff" was once regarded as a harmless eccentric at best, he or she is now seen as the mainstream of collegiate literary life. Stepping down to high school, junior high school or "middle school" levels, the phenomenon is even more striking. I recently taught an evening course at the College of Marin, in science fiction, and a high percentage of my students were librarians and English teachers from the high school and advanced elementary school level.

I asked the students why they were enrolled in my course, and again and again and again they told me the same story. "The kids today simply will not read *anything*—most of them. But the ones who *do* read don't want to read anything but science fiction. We're here to learn about science fiction so we'll know what books to recommend and to stock, which ones to avoid, and so on."

It's no surprise to state that the major leaders of large corporations are people who are willing and able to read. And if the only school kids today who are willing to read, read only science fiction . . . .

They read our stuff. It's up to us to provide them with books that point up real problems before they arrive, and that offer sound and humane responses to them. These leaders are hungry for ideas, and they are coming increasingly to us to get those ideas.

As for more direct and elaborate collaboration between science fiction writers and corporate leadership—I'm not sure that that's really in the cards. I certainly don't know how to go about setting it up. We can plant ideas, but the actual decisions of corporate leaders are generally based on hard-headed factual considerations, mainly economic considerations. At least, that was my observation of five years at Sperry Rand and seven years at IBM.

LARRY NIVEN

It isn't to be expected that writers will work well with other people. It can happen, but don't expect it. Science fiction writers in particular live in worlds of their own. Even those of us who collaborate, demonstrating an ability to work closely with *one* other human being, can't necessarily work with a corporation.

Remember the first season of *Star Trek*? Several science fiction writers of known competence were persuaded to write scripts. Their shows were the best of the lot. But the writers then dropped out. Even the money (fantastic!) wasn't enough to compensate for the drawbacks: other people of lesser talent mucking with their precious prose.

So the relationship will be loose.

First, corporations can make serious efforts to keep the science fiction field informed of what they're doing. This could be done in cooperation with the Science Fiction Writers of America. Or, Jerry Pournelle has been acting as liaison with the Jet Propulsion Laboratories (JPL) and other outfits. (JPL saw to it that a bunch of us watched the Voyager encounters with Jupiter. In contrast, NASA saw to it that a science fiction writer had to demonstrate that he was a newsman if he was to watch a Saturn Moon rocket launching. I've always thought it should have been the other way around: a newsman should have had to show at least one published science fiction story . . .)

We do our damndest to predict a probable future, problems and all. The corporations are shaping our futures. They can tell us how. Hell, they've all got publicity departments.

Second, our part is to write. Today's world society is as future-oriented as any that has ever existed (barring those whose concern was with life after death). Science fiction must keep us that way. Even bad science fiction, even anti-technology science fiction carries the basic message: that times change; that future generations will not think as we do and will face problems different from ours.

Don't even consider our obligations to the corporations for their efforts. Science fiction writers don't take orders worth a damn. If that seems unfair to the corporations . . . well, perhaps it is.

Third, the reverse? Corporation personnel may feel free to read science fiction; to learn from cautionary tales; to fall in love with a lunar mining scheme, and to try to make it come real. Robert Heinlein was writing about mass drivers (called "linear accelerators" or "electromagnetic cannons" then) many decades ago. Arthur Clarke wrote of communications satellites in geosynchronous orbit . . . .

This too is a loose relationship. We don't expect a corporation to take orders from us. But we'd love to have a hand in shaping a better future.

CHARLES SHEFFIELD

The question you pose is an interesting one, and I think that my feelings on it represent a minority viewpoint among science fiction writers. Let me first give a very brief answer, then amplify it later in my response. Ninety-five percent of science fiction writers have little or nothing to offer America's major corporations, either in coping with societal change or in planning for a more future-oriented society. Of the five hundred and odd members of the Science Fiction Writers of America, only a handful are going to be helpful to industry in the planning and adaptation process.

To explain why this is so, let me first quote from a passage written by Norman Spinrad. It is taken from his book, *The Star Spangled Future*. Spinrad writes: "The worlds of science fiction all too seldom can be comprehended as projections of vectors we're travelling in our own times. They all take place on the other side of some Great Discontinuity, be it a million years in time, or parsecs of space, or the great atomic war, or a Velikovsky two-cushion shot. Here the agonies and problems of our own times have been erased and cartoon heroes play out television scenarios. Dystopian warnings or logical positivist space opera, neither of them connects up to history.''

If we take a look at the science fiction that is appearing on the book shelves these days in ever increasing quantities, we find that Spinrad's point is on the mark. Many of the books are outright fantasy, even when they are labelled science fiction—a planet where the inhabitants all act out roles from the plays of Shakespeare must be considered a little unlikely.

Other books, ostensibly set in a possible future, are set so far out in time that there is no way they can be related to the process by which they arose from our present period. Many of them offer a society with the cultural habits of 1980 America, spread through the Solar System (or even the Galaxy). Still others are set on planets with alien species as the main characters. Some offer a bizarre mixture of human and alien life forms, interacting as comfortably as a couple of Manhattan business men in a mid-town bar—no culture shock, no language problems, and no xenophobia.

One book in five hundred, and perhaps one short story in a thousand, makes a conscientious attempt to extrapolate today realistically, and show an honest view of tomorrow. Why is the number so low? Two reasons: first, it is damned hard to grow a future organically from today's complex scene. Second, there is little sign that most readers really want realism. They want entertainment, and that often means cute aliens and fairy stories of other worlds and distant times.

There is a third reason that applies more to the writers than the readers. That is, relatively few science fiction writers have good grounding in the complicated mixture of politics, economics, hard science, and sociology that future projection requires. Let me stick my neck out a long way, and tell you who, in my opinion, has the background to be useful to the industrial groups who will study the real future. There are less of them than you might think.

Of the older generation, the undoubted master of the process, on the basis of past performance, is Robert Heinlein. He has written prophetically of many subjects—sometimes he has been wrong, of course, but that's inevitable. I am not sure that, at age seventy-three, he would want to be involved in any actual work on projecting the future—he has been sick, and is enjoying some relaxation.

Second only to Heinlein is Arthur Clarke, who in addition to the specific invention of the concept of the communications satellites, also led the

47

way in much of the space program development logic, when he was Chairman of the British Interplanetary Society and later.

The other "Big Two" of the older generation, Frank Herbert and Isaac Asimov, have nothing to show in predictive power that remotely rivals Heinlein and Clarke. And I don't think you'll easily get Clarke here again from Sri Lanka—he says he has made his last trip to this country, although he is still only in his early sixties.

Let's look now at the new writers, ones who should be picking up where Heinlein and Clarke are leaving off. This may be controversial, but I can tell you my prime candidates. The main point about these writers is that they have thought hard about the future, and the near future at that. Add in their strong imaginations, and you have a promising combination. I list them in alphabetical order, because I find it hard to list any preferred grouping. They are: Ben Bova, Joe Haldeman, Jerry Pournelle, Norman Spinrad, and Harry Stine (who writes as "Lee Correy" for his science fiction work). To this short list, I would perhaps add Donald Kingsbury, Dean Ing and James Hogan, based on limited exposure to their works— you need to see a good deal of somebody's writing before you can really gauge their views of things.

You will notice that there are no women on the list, and I don't think I'm being sexist. I have seen little near-term realistic projection attempts made by women science fiction writers, and the best of them seem little interested in the process. There are also some obvious omissions on my list—no Larry Niven, no Poul Anderson, no Robert Silverberg. All of them have the right kind of imaginative force for the work, but that's not quite where I think their interests lie.

So much for the people. I know that's not what was explicitly asked for, but I think it most important to have a feel for the type of person I would recommend, before I say how they would be used. It would be a disaster to use most science fiction writers in the role that I am now going to suggest.

Industrial groups are quite capable of hiring their own analysts and statisticians to plot numerical trends in prices, in productivity, in mortality from various diseases, in expenditures on defense and other areas of public use, and in every specific measure of American and international life. These same analysts are less aware, generally speaking, of the more subtle qualitative changes in attitudes that accompany, cause, and are caused by the numerical measures. Writers, on the other hand, are sensitive to social mores, and make it part of their work habits to observe the way that people live, spend money, talk, work, and play. That is the role in which they are best used, as part of a *corporate team*. Not as mystic gurus who come in for a day, pronounce on the future, and disappear again, but as an integrated part of a corporate long-range planning group, with responsibility to interpret the numbers that the other analysts are producing. The science fiction writers that I have listed are sufficiently well versed ín the sciences to comprehend statistical arguments, and to accept the idea that they will be creating a whole portfolio of futures, each with a different probability of being realized.

That's the way to use the select band of science fiction writers that I listed. Wire them into the planning groups, on a one, two and five-year basis, so that they can learn the business end, and the management can learn their thinking habits.

I can see only one possible problem with this idea. The writers in many cases have elected to follow the writers' life, mainly in order to *avoid* many of the more tedious elements of the business world. Luring them back in, unless it was clear that they would be listened to rather seriously, might not be at all easy.

ROBERT SILVERBERG

Science fiction writers aren't prophets by trade, and the specific views of the future that they depict in their stories are not to be taken as reliable indicators of events to come, or necessarily even of the writers' own views of what they think is ahead. Writers are artists before they're anything else, and even in science fiction idea and theoretical substructure are subordinate to narrative needs. (I often have written about things I think are pure fantasy, like time-travel, and about future developments that I don't really anticipate, such as the uncontrolled breeding in *The World Inside*.)

On the other hand, science fiction writers as a class are intelligent life-forms who tend to keep an eye on social trends, both out of sheer intellectual curiosity and for potential story material. Those of us who have had some material success at our trade also—at least I certainly do—monitor developments in the culture around us for the sake of preserving and enhancing our economic positions in a tricky and precarious environment. In short, we're bright people with one eye, at the very least, on the future, and although I don't recommend that the planners at major corporations spend a lot of time reading science fiction for news of tomorrow's society, I think very strongly that they can benefit from *talking* to science fiction writers about the shape of things to come. I've done some consulting work with corporations myself, on such matters as demographic change and current energy problems, and I think my views have been of some value to them.

A. E. VAN VOGT

My observation is that science fiction writers vary widely in their political and social philosophies. It is also noticeable that many of them started out as radicals and gradually made that famous age shift towards conservatism.

Obviously, a writer who is still in the liberal left period of his "development" will deal with American corporations only for money and not because the idea of such a relationship enthralls him. And the fact is that writers, composers, artists, etc., had their works stolen by enterprising publishers until copyright laws were finally enacted or unionization of certain crafts provided protection that was not available in the nature of business men.

By the way, none of this ever disturbed me. I looked at this situation in my systematic fashion. After all, I arrived at adulthood after Marxism—as an example—had had a good chance to prove once again that basic human nature does not change. Once more we have had a chance to observe those in power grab the best homes; we have watched them travel first class; we have noticed that they drive fine foreign cars, and we have seen that their children get the best jobs. Etc., etc., etc. Shall we say *ad nauseam*? No. Because this is timeless human nature at work.

What is sad is the pretense in the so-called socialist countries that human nature is *not* like this. Or, perhaps, that the time will come when it is not like this. (After all, Marx envisioned that, after a period of transition, there would be *no* government needed. Everybody would be honest all the time, minus police, minus army.)

We need to be thankful that the founders of America had no such illusions. They set up a system whereby the individual was free within the framework of laws—which merely forbade him to use force in anything that he did in relation to another human being's body or possessions. It subsequently took time to modify those laws slightly so that writers, workers, and people in general were protected from signing over all their rights because they were desperate.

In my early days I sold all rights to my stories in exchange for one cent a word. Later, the law of supply and demand worked in my favor. This happened when Horace Gold, as editor of *Galaxy*, began to buy only world and American serial rights. Evidently, my original publisher began to feel threatened by this, for suddenly the firm adopted a policy whereby, on request, I could get a reversion of all the rights they had previously bought except serial rights. Writers who had moved and therefore did not receive notice of the availability of rights, and did not request them, later—when the firm was sold—discovered that said rights were no longer available.

This unchanging human nature is what we have to deal with . . . in contracts, in ideas we present to business people, in all professions. A case in point: I discovered that certain publishers pretended that books were not out of print long after they had vanished forever from the market. I also discovered that some paperback publishers failed to pay promised royalities. What did I do? I devised a system whereby from those publishers I asked a larger advance on a straight term contract; and when that term expired simply asked for another advance or sold the book elsewhere.

I was, of course, not in the position of poor Mozart, who wrote all that marvelous music, and usually had to sell each composition outright because there was no copyright protection. I owned my rights for twenty-eight years, and had the right (in the twenty-seventh year) to renew for another twenty-eight. Was this a valuable right? It sure was. Because when I sold a story to a television show the legal department of the buyer requested proof that I still owned the copyright. (Presumably, if they had discovered that copyright had expired they would have ignored me totally.)

Later, I tried to help publishers who had gotten hold of certain of my titles on a long term basis . . . tried to help them sell those rights. Because, it is better to get fifty percent of something than fifty percent of nothing—and in those early years they were not qualified to sell rights that they controlled.

So the problem of dealing with large American corporations is multi-faceted. It is a problem which, first of all, inventors have confronted for a long time. But the fact is that American corporations are, essentially, all we have; and they are better than the self-deluded public corporations of the so-called socialist countries, in that the "almighty dollar" is what counts and not also the private opinions and philosophy of the person they deal with. (Though I'm sure that a well-known Communist or outspoken opponent would never be hired by a big firm—but at least he isn't barred from going elsewhere.)

Taking into account all of the foregoing we assume that a writer in dealing with a large American corporation reads his contracts carefully, has a time limit on everything, and—once involved—does not hold back, but gives what he has.

Within this frame, it is very likely that science fiction writers have a lot to give to corporations needing original ideas. My own situation is that I'm slow but observant. It takes a while for me to take account of all the factors. First of all I immerse myself thoroughly in what-is-the-current-situation and how-is-everything-done-now. I may come up with inter-mediate thoughts the first year but for the big stuff several years go by—at which time come a major original idea.

The reason that any science fiction writer is probably useful for long range planning and concepts is that, first, in reading science fiction, and

51

then writing it he has considered options for the future in a way that is different from all those people who are not, or have never been interested in science fiction.

What makes this important is that we are now in the second last decade leading up to the year 2000 A.D. Every year things move more rapidly. That's because the base of operations is so large. It's the inverted triangle effect. Technology is approaching the double millennium on a vast base of accomplishment; and soon we shall be off this small atom of a planet and out into the great universe.

We are about to be born in a meaningful sense of that word. It's like the ancient ape coming down from the tree and experiencing an expansion of consciousness as a result. The expansion of consciousness that will result from the average man going out into space is almost impossible to visualize. Try to imagine those first tree ancestors of ours having the thought that someday a descendant of theirs would be a pilot of a super-jet 747.

Very likely the first people from science fiction that will be asked by Big Business to act as idea men will be those with degrees in physics, chemistry, and other physical sciences. Theoretically, they are better qualified to come up with sound sense. The statistics on that can be tabulated later.

In the long run I really don't believe there will be much difference. The giant human brain is detectable in every science fiction writer I have met so far; and so, since all of us have a large "feel" for science, it's really just a matter of an older person doing in a more concentrated way what those with the degrees did in their nonage: get the simple facts straight; and go on from there.

Of course, there's another problem. When I was asked by a computer corporation executive last year if I would accept employment as an idea man—if it were offered me—my first reaction was no. One of the automatic reasons was that I doubt if they would offer me enough pay. My expenses these days are huge. Among other activities, I am looking for the two hundred languages and dialects that I believe are available in the Los Angeles area, and I am recording them at my own expense at a professional recording studio, and paying the people who speak the various languages for their time in preparing and recording for me. Since I am doing this by a new system for learning—which I devised after looking over the problem of language learning in this country (an original idea that took a long time to jell—though it's basically so simple, no wonder the others missed it).

I'm also considering the problem of prolonging life and energy—after all I shall be sixty-eight this year.

I notice that other science fiction writers are also involved in various endeavors, and are also living high; so we have a problem of availability.

It is my belief that a new book, *The Evolving Brain*, by Tony Busan explains why science fiction writers are automatically engaged in personal activities involving the real world. It appears that the right side of the brain has to do with art, music, and fantasizing; and that the more you exercise the fantasizing science fiction part of that right lobe the stronger becomes a corresponding activity in the reality-language-logic-mathematical left lobe.

The language aspect and several other activities happened automatically with me before I ever read the book—which has just been published by Holt, Rinehart & Winston.

It would be a mistake for a science fiction writer to abandon writing for a full-time job as an idea man—since he should continue strengthening the reality side by working with fantasy, or music, or art *while* he works at the reality job of thinking up ideas of value in the world of science, technology and business.

Looks like we all have a fine balancing job to do with the equipment we have in our heads. And it could be that by actually writing more fully worked out future options we can make our contribution without ever tying ourselves down to a salary.

But the contribution has to be in the idea area; so it seems to me.

JOHN VARLEY

I don't really have much useful to say about the question you pose. For one thing, I don't think I can speak for other science fiction writers; many of them are sure to be eager for such a role. I can only speak for myself, and I recognize myself as only a storyteller and no kind of prophet at all. If I tried to tell major corporations or anyone at all what was coming in the future and how to plan for it and how to prevent disasters and how to become more future oriented, I would be doing nothing more than indulging in false self-importance. (I suspect a lot of the answers you get will be doing the same thing, but will at least *sound* better than anything I could say.)

My attitudes about the future are wrapped in extreme pessimism, I don't really believe anybody can do anything. My impulse, since you asked what *this* science fiction writer might do in working with the major corporations, would be simply to tell them to stop. Stop everything you are doing now . . . everything. Think it all out for a few decades, and then start back in slowly when you think you might know what the mistakes were. I do not propose this as a rational solution. My thinking about the future is anything but rational.

JACK WILLIAMSON

I do think science fiction writers have a role to play in working with nearly anybody to cope with societal change and plan for a more future-oriented society.

Science fiction writers have occasionally made sound predictions about the future, and I think there is a significant relationship between science fiction and futurology. H. G. Wells, I believe, was the principal shaper of modern science fiction. I believe he was also the inventor of futurology (see his lecture and book, *The Discovery of the Future*, published in 1902). I think his career in both fields was largely determined by his year with Thomas Henry Huxley as a biology student. He learned Darwinian evolution, and he was a pioneer in applying it both in fiction and in the real world. In fiction, *The Time Machine* projects human evolution into the future; *The Island of Dr. Moreau* is a parable of evolution; *The First Men in the Moon* is a somewhat sardonic look at genetic engineering. In fact, he became more and more concerned with the alarming shapes of the actual future as he foresaw it, and he spent most of his life as a practicing futurologist and as a campaigner for the world state that he came to see as the alternative to world destruction.

Yet, very little science fiction is written as simply prediction. For the sake of drama, the writer generally opts for the unlikely disaster instead of the more likely human triumph.

The important thing that science fiction can do is to condition our culture to accept the fact of change. We tend to fall in love with the status quo. All sorts of individuals and organizations gain a stake in it, battle to defend it, and look with dread at any event that might change it.

In fact, the world is changing under the impact of technology. Some of the changes are already history, of course; many more are obvious; vaster changes to come are hardly suspected. The computer, for example, seems certain to transform our culture profoundly—perhaps as profoundly as the invention of fire or the invention of agriculture. Science fiction has looked at some of the possibilities, but hardly enough to reveal the actual shape of things to come.

For one obvious example of the way we cling to the status quo, I might mention the current debates about nuclear energy. Actually, it is—I'm pretty certain—the safest and cheapest power we can use for the rest of this century. The need is so desperate that I can't believe we'll turn it off. But if you listen to Jane Fonda and Barry Commoner . . . .

I would expect science fiction writers to have more influence indirectly than directly on the major corporations. A few of them—some of the biggest names in the field—have been employed to speak at board meetings and conventions and to write for house organs. But I think most corporate planning ignores science fiction. We haven't entirely recovered from our generation in the pulp ghetto, and I suspect that most corporations, like most other influential or powerful institutions, are still dominated by C. P. Snow's "traditional academic culture," still ignorant or suspicious of the "culture of science." I hope science fiction can build a bridge between the two cultures. That process, I think, has already begun—but a look at the headlines will show that it still has some way to go.

ROBERT ANTON WILSON

It seems to me that corporations (and governments also) will increasingly need to hire futurists as consultants, as routinely as they now do with other social scientists. That is, futurism—the art and science of extrapolating *alternative* future scenarios—will become one of the most important social sciences for all persons in decision-making positions.

Science fiction writers will play a role in this because there is a continuous feedback between science fiction and future studies. Most futurists I know read science fiction and are influenced by it; and most science fiction writers read futurist books and magazines and are influenced by them. In addition to this, many science fiction writers are also active futurists.

Science fiction, because it is a form of literature, "fleshes out" future scenarios in a way that the best expository science fact writing simply cannot do. For instance, any corporate body trying to comprehend what longevity means would learn many important things from recent science fiction. They could learn, say, the state of current research and what can be expected in the next twenty to thirty years from a factual book like Rosenfeld's *Prolongevity*; but to get a feel for the social and psychological consequences of longevity, they would do better to consult Harrington's novel, *Paradise One.*

In other words, if a corporation asked me right now to recommend a panel to brief their executives on the longevity revolution, I would recommend not just several leading researchers, such as Drs. Segall, Stehler, Comfort, etc., but also several science fiction writers who have vividly imagined the social upheavals that might result, such as Harrington, Heinlein, Simak, and Bester.

CONTRIBUTORS

POUL ANDERSON. Born in Bristol, Pennsylvania, he attended the University of Minnesota, where he received his B.A. in Physics. Since becoming a full-time writer, he has worked in numerous genres, among them science fiction, fantasy, mystery, poetry, and mainstream. As a science fiction writer, he has received numerous honors, including the Hugo, Nebula, and Forry awards. His first professional sale, "Tomorrow's Children," appeared in *Astounding*, in 1947. According to English writer, journalist, and editor, Brian Ash, his popularity is partly attributable "to the versatility and sheer volume of his work, coupled with a realistic approach and generally straightforward literary style." The author of such popular works as *Trader to the Stars, A Midsummer Tempest, Brain Wave, Beyond the Beyond, The Star Fox* and *Tau Zero*, he has never been content to rest on his laurels. Indeed, he has continuously set a higher and higher standard for himself. In the process, he has eschewed the lure of the formula novel, preferring instead to experiment with new ideas, new techniques, new approaches. He lives in Orinda, California.

MILDRED DOWNEY BROXON. Convent-educated, she attended Seattle University, where she earned a B.A. in Psychology and a B.S. in Nursing. She worked for a time as a psychiatric nurse at Harborview Hospital in Seattle, dealing with violent psychotics on the unit's locked ward. She became "addicted" to science fiction at the age of eight, because of an interest in astronomy and space travel. Since then, science fiction has remained a "ruling passion" in her life, and she sold her first story in 1972. She is a member of the Science Fiction Writers of America, whose *Handbook* she edited, and in which she served two terms as Vice-President. Aside from fandom, she is active in the Society for Creative Anachronism, a group dedicated to recreating the Middle Ages as they should have been. Other interests include herpetology, Egyptology, Irish history and mythology, languages, and cats.

OCTAVIA E. BUTLER. The author of such well-received works as *Patternmaster, Mind of My Mind, Survivor,* and *Kindred,* she describes herself as "a hermit—living in the middle of Los Angeles—a writer of both science fiction and mainstream fiction, a pessimist if I'm not careful, a feminist, a black, a quiet egoist, a former Baptist, an oil-and-water combination of ambition, laziness, insecurity, certainty, and drive." A self-made, self-taught writer, she pursues her craft with enthusiasm and dedication. As a youngster, she never envisaged that it would be possible for a black person, like herself, to escape the boredom, the dreariness, and the resignation which accompanied her life. To compensate, she turned to reading, specifically science fiction, in the hope of discovering worlds which promised greater joy, excitement, and adventure. The library became her second home. Soon, reading was no longer sufficient to quench her appetite. She turned to writing as a means of self-expression. But even then, she had no idea that one could make a living as a writer, which was just as well, she admits, since it was "a time to love writing before I knew it was supposed to be work."

C. J. CHERRYH. Born in St. Louis, Missouri, she attended Johns Hopkins University, where she received her B.A. in Latin and her M.A. in Classics. One of science fiction's most respected talents, she has been writing since the age of ten. After a string of rejections, she finally cracked the commercial market, with the sale of her first novel, *Gate of Ivrel,* in 1976. Since then, she has authored several popular works, including *Brothers of Earth, Hunter of Worlds, The Faded Sun, Well of Shiuan,* and *Fires of Azeroth.* Known for her penetrating insights and deft craftmanship, she has played a pivotal role in defining the genre and extending its perimeters. Asked why she writes science fiction, she observes: "It's a genre that pushes at the outer frontiers of human knowledge, pushes at the limits of our understanding about the universe, asks why and tries to give an answer to it. I don't think you could ask more of a literature." She resides in Oklahoma City, Oklahoma.

GORDON R. DICKSON. A Nebula and Hugo award-winning author, he was born in Edmonton, Alberta, Canada. He attended the University of

Minnesota, where he earned a B.A. in English. During that period, he met Poul Anderson, who attended the same institution, and with whom he would later collaborate on several novels. A prodigious writer, his fiction has appeared in most of the field's leading publications. He is the author of numerous works, among them *Necromancer, Tactics of Mistake, Soldier, Ask Not, The Genetic General, Three to Dorsai, Time Storm, Home From the Shore,* and *The Spirit of Dorsai.* He has also produced several collaborations, chief of which are *Earthman's Burden* (with Poul Anderson), *Planet Run* (Keith Laumer), *Gremlins, Go Home!* (with Ben Bova), and *Lifeship* (with Harry Harrison). A former President of the Science Fiction Writers of America, he lives in Minneapolis, Minnesota.

RAYMOND Z. GALLUN. Born in Beaver Dam, Wisconsin, he attended the University of Wisconsin, Alliance Francaise, and San Marcos University. A pioneer in the establishment of modern science fiction, he played a salient role, in the words of writer-critic John J. Pierce, in "setting in motion the evolution of science fiction from crude pulp fiction to a form increasingly imaginative and literate." He launched his career in 1929, with the sale of "The Crystal Ray," to *Air Wonder Stories.* However, it was with the publication of his "Old Faithful" series that he first won popular attention. As an active writer, he was known for his inventive ideas, his philosophical imagination, his skill at characterization, and his mastery of technical detail. His novels include *The Planet Strappers, People Minus X,* and *The Eden Cycle.*

JAMES E. GUNN. A Professor of English and Journalism at the University of Kansas, Lawrence, he specializes in fiction writing and science fiction. He has written screenplays, radio scripts, verse, scholarly articles, and criticism. However, most of his work has been in the field of science fiction. During his long career, he has published more than seventy-five stories in magazines and books; most of them have been reprinted, some as many as a dozen times. He is the author of seventeen books and the editor of five; his M.A. thesis (about science fiction) was serialized in a pulp magazine. Four of his stories were dramatized over NBC radio, one ("The Cave of Night") was dramatized on television's "Desilu Playhouse," and *The Immortals* was dramatized as an ABC-TV "Movie of the Week" and became an hour-long series, "The Immortal." He is best-known for such works as *Star Bridge* (with Jack Williamson), *The Joy Makers, The Immortals, The Listeners, The Magicians,* and *The Road to Science Fiction.*

ISIDORE HAIBLUM. A Yiddish novelist and humorist, he was born in Brooklyn, New York. As a child, he grew up speaking Yiddish, and has been speaking, reading, and writing about Yiddish ever since. He has published numerous articles and book reviews on various Yiddish and non-Yiddish topics, in magazines such as *Midstream, Moment,* and

The National Jewish Monthly. He attended the City College of New York, graduating with a B.A. in English and Social Science. There, he studied Yiddish with the world-renowned linguist, Dr. Max Weinreich, and won honors in Yiddish language and literature. After a series of uneventful jobs, he developed an interest in science fiction, which, he says, changed his life. He is the author of a number of novels, among them *The Tsaddik of the Seven Wonders, Transfer to Yesterday, The Return, The Wilk Are Among Us*, and *Interworld*. He lives in New York City.

JAMES P. HOGAN. Born in London, England, he was educated at the Cardinal Vaughan Grammar School, Kensington. Later, he studied General Engineering at the Royal Aircraft Establishment, Farnborrough, specializing in electronics and digital systems. After working as a systems design engineer, he transferred into selling and joined the computer industry as a salesman. He recently quit his job at Digital Equipment Corporation as a Sales Consultant, where he specialized in the applications of mini-computers in science and research, to write on a full-time basis. A master of "hard" science fiction, his novels include *Inherit the Stars, The Genesis Machine, The Gentle Giants of Ganymede, Two Faces of Tomorrow*, and *Thrice Upon a Time*. Asked to discuss his approach to science fiction, he explains: "I try to use science fiction as a means of contributing something to reinforcing society's confidence in science and in itself as the only viable way of ultimately solving its problems. I like to see plausible science, positive thinking, and optimistic pictures of the human race facing its problems and solving them instead of running away, fearing technology, and relying on magic, mysticism, pseudo-science, or the like."

ROBERT A. W. "DOC" LOWNDES. A member of the well-known Futurian Society, he achieved prominence in the publishing field, editing such early magazines as *Future Fiction, Science Fiction Quarterly, Dynamic Science Fiction*, and *Original Science Stories*. In addition, he wrote *The Duplicated Man* (with James Blish), *The Mystery of the Third Mine, The Puzzle Planet*, and *Believers' World*. His literary essays, published in *Famous Science Fiction*, have been collected in a work titled *Three Faces of Science Fiction*. He is currently the Production Chief for *Luz* and the Production Associate for *Radio Electronics*. An ardent fan of murder mysteries, he is an investitured member of the Baker Street Irregulars (the central Sherlock Holmes society of the U.S.). He still writes occasionally for the fan magazines, as well as for the leading Sherlockian and mystery lovers publications. He resides in Hoboken, New Jersey.

RICHARD A. LUPOFF. A science fiction fan at the age of five, he was "hooked" after reading *Maximo the Amazing Superman*, by Russell R. Winterbotham. His own writing career began with sports reporting for newspapers in New York, New Jersey, and Pennsylvania. A leading science fiction fan, he won the Hugo award for the fanzine, *Xero*, which he co-edited with his wife, Patricia. In the early 1960s, he served as editor of Canaveral Press, publisher of books by Edgar Rice Burroughs, E. E. Smith, and L. Sprague and Catherine de Camp. His own first novel, *One Million Centuries*, was published in 1967. In recent years, he has written many popular short stories and novels, including *The Triune Man, Sword of the Demon*, and *Space War Blues*. His non-fiction works include *Edgar Rice Burroughs: Master of Adventure, All in Color for a Dime,* and *The Comic-Book Book*. He has also served as writer and consultant for a number of television series and theatrical films produced by Twentieth Century-Fox and Paramount Pictures. He lives in Berkeley, California.

LARRY NIVEN. Born in Los Angeles, California, he graduated from Washburn University, with a B.A. in Mathematics. His first sale was "The Coldest Place," to *Worlds of If*, in 1964. Since then, he has become one of science fiction's most popular writers, owing to the success of such books as *World of Ptaavs, A Gift From Earth, Neutron Star, Ringworld, The Shape of Space, The Protector,* and *The Flight of the Horse*. His recent collaborations with author Jerry Pournelle—*Inferno, The Mote in God's Eye,* and *Lucifer's Hammer*—have catapulted him into national prominence. A master of "hard" science fiction, he has won virtually every honor the genre has to bestow, including the Hugo, Nebula, Ditmar, and E. E. Smith awards for his fiction, as well as his service to the field. Asked about his growing popularity, he opines: "I had a bit of good luck. The New Wave hit its stride just as I was starting my career. New Wave stories typically concentrate on experimental styles and the exploration of character, to the detriment of extrapolation, solid background and storytelling. It's a seductive approach, a fine excuse for bad writing and not doing one's homework. All the new writers were writing New Wave at the time except me. My only competition came from Robert Heinlein, Poul Anderson, Hal Clement and the like, the very people I wanted for my peers." In addition to writing, he enjoys sailing, backpacking, poker, Scrabble, fandom, and Georgette Heyer literature.

CHARLES SHEFFIELD. Author of such popular works as *Sight of Proteus* and *The Web Between the Worlds*, he is Vice-President of Earth Satellite Corporation and First Vice-President of the American Astronautical Society. Born and educated in England, he immigrated to America in 1962 and joined Computer Usage Company, one of the first independent computer software companies. His involvement with the space program began the following year, with the Navy's satellite geodesy program, and has continued ever since. He has served as a consultant to NASA on the computational implications of earth resources satellites, and later was involved in the development of wholly digital methods for reduction and information extraction of earth resources data. He holds advanced degrees in mathematics and theoretical physics, and has published nearly fifty scientific papers on a variety of subjects, including nuclear physics, orbit computation, earth resources, large-scale computer systems, and general relativity. He resides in Bethesda, Maryland.

ROBERT SILVERBERG. A full-time, free-lance writer since 1953, he sold his first article, "Fanmag," to *Science Fiction Adventures*. His first story, "Gorgon Planet," appeared the following year in *Nebula Science Fiction*. To date, he has authored more than seventy science fiction books and over two hundred uncollected short stories. He has produced some sixty non-fiction works and has written extensively for periodicals of many different kinds, including numerous non-science fiction publications. He has also edited more than forty anthologies. The author of many important novels, he is perhaps best-known for such works as *Tower of Glass, Downward to the Earth, Son of Man, Dying Inside, To Live Again, Nightwings, Shadrach in the Furnace,* and *A Time of Changes.* In 1974, he did the unthinkable. After a successful twenty-year career in science fiction, he quit. He felt ignored, disappointed over his failure to write that all-important "big book" that would bring him mainstream recognition. Today, he is writing again. After a half-decade hiatus, he recently completed an epic adventure titled *Lord Valentine's Castle,* which has now been published to great acclaim. Asked whether science fiction still excites him, he observes: "Yes, very much so, quite the way it did when I was nine years old. It shows me things I couldn't see otherwise. I don't go to science fiction for social satire. I don't go to science fiction for criticism of society. I can get that by just looking at a newspaper and muttering to myself. I go to science fiction for sweep, for vision, for beauty, for the mother-ship hanging overhead, for those special moments of wonder. Admittedly, it's harder and harder to find them. It's much easier to have your mind blown when you're nine years old. But I can still find those moments occasionally. It's in that hope of discovery that I keep going back. And I still find that magic, much to my amazement and pleasure."

JACK VANCE. Born in San Francisco, California, he received his B.A. in Journalism at the University of California, Berkeley. He broke into the science fiction field in 1945, with the sale of "The World-Thinker" to *Thrilling Wonder Stories.* His first novel, *The Dying Earth,* followed five years later. Since becoming a full-time writer, he has published a myriad of books, among them *The Dragon Masters, The Blue World, Emphyrio, The Anome, The Dirdir, The Pnume, The Palace of Love,* and *Showboat World.* Drawing on several disciplines for inspiration— anthropology, sociology, political science, psychology, music, art, etc.— his fiction is rich in ideas—immortality, linguistics, genetic engineering, communication, alien life, and numerous others. An award-winning writer, he has won the Nebula, Hugo, and Jupiter for his literary efforts.

Still, despite his achievements, he is a very private man. According to English editor and critic, Malcolm Edwards: "He is renowned for his reticence concerning himself and his stories, maintaining such a low profile that a rumor, which started in 1950, that he was another Henry Kuttner pseudonym, was still perpetuated in some quarters twenty years later, notwithstanding Kuttner's death in 1958."

A. E. VAN VOGT. Author of more than fifty books and two hundred short stories, he stands behind Robert Heinlein, Isaac Asimov, and Arthur Clarke as one of the field's most respected writers. Dubbed "the undisputed Idea Man of the Futuristic Field," by Forrest Ackerman, he has delved into such cerebral concepts as hypnotism, telepathy, semantics, "similarization," and Dianetics. His delight in intellectual adventure permeates the author's real life as well. During the course of his career, he has explored the dynamics of violence, pioneered a technique for recording dreams, and is presently attempting to simultaneously master two hundred world languages. His analytical approach is reflected in such classic works as *Slan, The World of Null-A, The Voyage of the Space Beagle, Destination Universe!, The Weapon Shops of Isher, The Darkness on Diamondia*, and *Children of Tomorrow*. His fiction has been translated into several languages, among them French, German, and Italian, and even recorded on "talking records" for the blind. Yet, despite his success, he still doggedly yearns to crack new systems ; develop new insights, new patterns of behavior. According to the author: "I am strictly a systems person. I've only gradually extended myself into the world, primarily as a result of adding new thought systems to my repertoire. If I don't have a system for something, I will remind myself of that fact and keep looking at the blank area in amazement. One day an insight will flash at me and I'll have the problem solved."

JOHN VARLEY. The winner of the 1979 Nebula award for his novella, "The Persistence of Vision," he is one of science fiction's most popular writers, owing in large measure to the success of his bestselling novel, *The Ophiuchi Hotline*. He recently published his second book, *Titan*, which has also met with considerable success. He began writing science fiction in 1973, primarily for financial reasons. SF, along with film, were lifelong personal interests. After failing to place his first novel, *Gas Giant*, he discovered a guest editorial by Robert Heinlein, in the January, 1974 issue of *Analog*, which served to point him in the right direction. His first-sale story appeared later that year in *Fantasy & Science Fiction*. Since then, his fiction has appeared in nearly every major science fiction publication.

JOAN D. VINGE. The author attributes her early fascination with space to a small telescope her father had in the back yard, which they used on summer nights to look at the moon and planets. Later, she began to write stories (usually about horses), which she often illustrated. When she was fifteen, she stumbled on her first science fiction novel, *Storm Over Warlock*, by Andre Norton. She was "hooked," and from then on read almost nothing but science fiction and fantasy. In college, she wandered through six majors (all unofficially), until she found archeology-anthropology. Once again, she credits Andre Norton and her book, *The Time Traders*, for her interest. As with science fiction, all it took was one and she was "hooked." Upon graduation, she wrote her first serious science fiction story, "Tin Solder." Encouraged by her then husband, Vernor Vinge (also a SF writer), she sent it out, and after a couple of fits and starts, Damon Knight bought it for *Orbit 14*. She was "hooked" again, and since then she has been, with more and more dedication, a full-time writer of science fiction. To date, she has written several novelettes ("Eyes of Amber," "Media Man," "Tin Soldier," "To Bell the Cat"), a number of novellas ("The Crystal Ship," "Legacy," "Fireship," "Mother and Child"), and two novels (*The Outcasts of Heaven Belt* and *The Snow Queen*).

JACK WILLIAMSON. Writing more or less steadily since his first sale in 1928, he has published more than three million words of magazine science fiction and thirty-odd books, with total sales of well above two million copies. His efforts have won him numerous honors, including the Science Fiction Hall of Fame Award, the Pilgrim Award, and the Grand Master Nebula. He received his B.A. and M.A. from Eastern New Mexico University and his Ph.D. from the University of Colorado. Until his retirement, he taught English at Eastern New Mexico University, where he specialized in modern fiction, literary criticism, and linguistic theory. Known as a pioneer teacher of science fiction, he was active in its establishment as a legitimate academic subject. In this regard, he published a descriptive list of some five hundred college-level courses, *Teaching Science Fiction*. He is best-known for such books as *The Legion of Space, Darker Than You Think, The Humanoids, Seetee Ship, Star Bridge* (with James Gunn), *The Reefs of Space* (with Fredrik Pohl), *The Pandora Effect, Undersea City* (with Frederik Pohl), and *H. G. Wells: Critic of Progress*. A former President of the Science Fiction Writers of America, he resides in Portales, New Mexico.

ROBERT ANTON WILSON. A former editor of *Playboy* magazine, he has been writing and lecturing on the future for many years. With nine books and over two thousand articles to his credit, he is the co-author (with Bob Shea) of *Illuminatus!*, which has been widely acclaimed as a ground-breaking literary work in the areas of science fiction and political satire. His recent effort, *Cosmic Trigger: The Final Secret of the Illuminati*, is the non-fiction sequel to his three novels on the subject, and deals with merging trends in physics and parapsychology. In addition to writing, he serves as President of the Exo-Psychology Institute,

Director of the Prometheus Society, and Vice-President of the Institute for the Study of the Human Future. Asked whether he enjoys writing, he exclaims: "It's sheer pleasure. I'm definitely a style-oriented writer. Every paragraph is a challenge, and when I get the paragraph organized just the right way, I experience a great sense of bliss, such as a mathematician experiences when he solves a difficult equation. It's a head game, a brain exercise, and it's lots of fun if one's attuned to that type of mental exercise. Every writing project is a growth project, especially if one has the aim, as I have, of never repeating myself. I keep trying to do things I've never done before, which means that every writing job entails another brain change operation."

CHELSEA QUINN YARBRO. Born in Berkeley, California, she attended Bay Area public schools and studied at San Francisco State University. She has worked as a theater manager and playwright, as a counselor for mentally disturbed children, as a statistical demographic cartographer, and as a tarot card and palm reader. In addition to writing, she is a serious composer and an occultist. She has sold twenty-seven short works of fiction and fifteen books, including *Ogilvie, Tallant & Moon, Time of the Fourth Horseman, Hotel Transylvania,* and *False Dawn.* A former Secretary of the Science Fiction Writers of America, she lives in Berkely, California.

ROGER ZELAZNY. He began writing professionally in 1962, and has won both the Hugo and Nebula awards. He is the author of approximately eighty-five short stories and articles. His fiction has appeared in numerous anthologies and magazines, including *Amazing Stories, Analog, Asimov's SF Adventure Magazine, Destinies, Galaxy, Magazine of Fantasy & Science Fiction, New Worlds,* and *Unearth.* He has published twenty-five novels, among them *This Immortal, The Dream Master, Lord of Light, Damnation Alley, Nine Princes of Amber, The Guns of Avalon, Sign of the Unicorn,* and *The Courts of Chaos.* His book, *Damnation Alley,* was filmed by Twentieth Century-Fox under the direction of Jack Smight, with a cast headed by Jan-Michael Vincent, George Peppard, and Dominique Sanda. Various other novels are also under film option. He lives in Santa Fe, New Mexico.

www.ingramcontent.com/pod-product-compliance
Lightning Source LLC
Chambersburg PA
CBHW022131280326
41933CB00007B/647